Cutting Edge
STEM Careers

Cutting Edge Careers in
ENGINEERING

Carla Mooney

Central Islip Public Library
33 Hawthorne Avenue
Central Islip, NY 11722

ReferencePoint Press®

San Diego, CA

3 1800 00357 4619

About the Author

Carla Mooney is the author of many books for young adults and children. She lives in Pittsburgh, Pennsylvania, with her husband and three children.

© 2021 ReferencePoint Press, Inc.
Printed in the United States

For more information, contact:
ReferencePoint Press, Inc.
PO Box 27779
San Diego, CA 92198
www.ReferencePointPress.com

ALL RIGHTS RESERVED.
No part of this work covered by the copyright hereon may be reproduced or used in any form or by any means—graphic, electronic, or mechanical, including photocopying, recording, taping, web distribution, or information storage retrieval systems—without the written permission of the publisher.

Picture Credits:

Cover: Cineberg/Shutterstock
 6: Maury Aaseng
10: Jacob Lund/Shutterstock
27: SeventyFour/Shutterstock
42: Monkey Business Images/Shutterstock
60: Suwin/Shutterstock
67: Africa Studio/Shutterstock

LIBRARY OF CONGRESS CATALOGING-IN-PUBLICATION DATA

Name: Mooney, Carla, 1970– author.
Title: Cutting Edge Careers in Engineering/by Carla Mooney.
Description: San Diego, CA: ReferencePoint Press, Inc., [2021] | Series: Cutting Edge STEM Careers | Includes bibliographical references and index.
Identifiers: LCCN 2019051034 (print) | LCCN 2019051035 (ebook) | ISBN 9781682828670 (library binding) | ISBN 9781682828687 (ebook)
Subjects: LCSH: Engineering—Vocational guidance—Juvenile literature.
Classification: LCC TA157 .M587 2021 (print) | LCC TA157 (ebook) | DDC 620.0023—dc23
LC record available at https://lccn.loc.gov/2019051034
LC ebook record available at https://lccn.loc.gov/2019051035

Contents

An Innovative Career	4
Software Development Engineer	8
Nanotechnology Engineer	16
Biomedical Engineer	23
Robotics Engineer	31
Industrial Engineer	39
Environmental Engineering Technician	47
Renewable Energy Engineer	55
Modern Structural Engineer	63
Source Notes	71
Interview with a Robotics Engineer	74
Other Jobs in Engineering	77
Index	78

An Innovative Career

A round the world engineers are developing the cutting edge technologies and innovative materials of the future. At Toyota, engineer and material scientist Minjuan Zhang and her team are working to build a type of invisibility cloak for vehicles. Zhang's main research focus is studying how light interacts with materials. She and her team are investigating how to use lenses and polarized light to make the internal structures of a car appear invisible. By doing so, they hope to provide unobstructed views for future car operators. "We could still keep the same structures, but we could make them invisible so we could improve the view of the driver,"[1] Zhang explains.

At Carnegie Mellon University in Pittsburgh, Pennsylvania, Carmel Majidi, an associate professor of mechanical engineering, and his team of engineers are working to develop a method to allow machines that have suffered severe mechanical damage to repair themselves like living organisms. The material they are working on is made from liquid metal droplets suspended within a soft elastomer, a stretchable material made of chains of molecules. When damaged, the metal droplets rupture like blood in an animal. They form new connections with nearby droplets and reroute the machine's electrical signals. "Other research in soft electronics has resulted in materials that are elastic and deformable, but still vulnerable to mechanical damage that causes immediate electrical failure," explains Majidi. "The unprecedented level of functionality of our self-healing material can enable soft-matter electronics and machines to exhibit the extraordinary resilience of soft biological tissue and organisms."[2]

Engineers are pioneering this research. They are creating new materials, machines, and processes to improve life. Renewable energy engineers are designing machines to improve solar, wind, and other renewable energy sources. Biomedical engineers are creating 3-D organs for transplants and new tools for surgery. Nanotechnology engineers are studying materials at the molecular level to improve products from sunscreen to electronics. Structural engineers are using the latest research, materials, and technologies to make sure buildings, bridges, and other human-made structures are safe and secure. Engineers in every field are on the front lines of making the dreams of the future a reality.

Engineering Careers in Demand

To keep up with the world's ever-evolving technologies, employers around the world are demanding engineering graduates with advanced skills. According to a 2019 survey of best college majors for a lucrative career by *Kiplinger's Personal Finance* magazine, engineering degrees took seven spots in the top ten careers that come with the best hiring prospects and pay.

That means engineering professionals are in demand, and employers are willing to pay top dollar for them. According to a 2017 salary analysis across fifteen countries by the Hay Group division of employment advisory firm Korn Ferry, STEM careers were among the highest-paid careers in every country. For example, the study found that entry-level engineers in the United States could expect to make 19 percent more than the national average entry-level professional salary. "It's important to note that many factors go into determining salaries," says Bob Wesselkamper, Korn Ferry Hay Group global head of rewards and benefits solutions. "However, graduates who choose certain career paths where talent is in high demand, like engineering or technology, can expect to make more than their peers, regardless of the country in which they reside."[3]

On the Cutting Edge

For students who enjoy math and science and want to be on the cutting edge of research and development, a career in engineering can be a great fit. For many engineers, the fact that the job is constantly changing and pushing boundaries is appealing.

Attributes That Matter to Employers

Written communication skills and the ability to solve problems are at the top of the list of attributes employers look for when considering new hires. This is the finding of a report titled "Job Outlook 2019." The report comes from the National Association of Colleges and Employers (NACE), an organization that surveys employers nationwide to learn more about their hiring plans in connection with recent college graduates. Other desirable attributes include the ability to work in a team setting, showing initiative, analytical skills, and a strong work ethic.

Attribute	% of Respondents
Communication skills (written)	82.0%
Problem-solving skills	80.9%
Ability to work in a team	78.7%
Initiative	74.2%
Analytical/quantitative skills	71.9%
Strong work ethic	70.8%
Communication skills (verbal)	67.4%
Leadership	67.4%
Detail oriented	59.6%
Technical skills	59.6%
Flexibility/adaptability	58.4%
Computer skills	55.1%
Interpersonal skills (relates well to others)	52.8%
Organizational ability	43.8%
Strategic planning skills	38.2%
Tactfulness	25.8%
Creativity	23.6%
Friendly/outgoing personality	22.5%
Entrepreneurial skills/risk-taker	16.9%
Fluency in a foreign language	11.2%

Source: "Job Outlook 2019," NACE, November 2018.
www.odu.edu/content/dam/odu/offices/cmc/docs/nace/2019-nace-job-outlook-survey.pdf.

Louise Campion is a project engineer with RPS Group, a global professional services firm that defines, designs, and manages projects from energy to transportation. She says that one of the most interesting and challenging aspects of working as an engineer is the constant innovation and change she experiences on a daily basis. Campion explains:

> Modern engineering methods require openness towards innovation and encourage holistic approaches. . . . Engineering has diverse applications and the success of projects relies on multifaceted teams and collaborative input. As a professional in this field, I am not just an engineer I am a planner, a designer, a scientist, a social entrepreneur and an advocate for environmental sustainability. Engineering is a continually evolving sector; technology and governance move quickly and things are never stale in this profession—I rarely do the same thing day in, day out. The rate of change makes for a challenging and exciting career.[4]

As engineers like Campion attest, the field is diverse, with opportunities in almost every discipline, from computers to medicine to chemistry. Talented individuals can find their place in the variety of engineering jobs available and build the technologies of tomorrow.

Software Development Engineer

What Does a Software Development Engineer Do?

Who creates the computer programs and applications that run on computers, laptops, tablets, and mobile devices? That is the job of software development engineers. They are the creative force that dreams up, designs, and develops the computer programs used worldwide. Software development engineers build operating systems, business applications, mobile and web applications, computer games, networking systems, and more.

Software development engineers drive the development process for a new software program. To begin, they analyze the needs of users. What do users need the software to do? What features do they want in the software? What creates a smooth user experience? Software development engineers determine the software's core functionality. They understand other user requirements such as security and performance features. Armed with this information, software development engineers design and develop software that meets users' needs. They design each piece of an application or system and plan how the pieces will fit together as a whole. Then they create models and diagrams for computer programmers that show what code will be needed to build the application. In some companies,

---

A Few Facts

Number of Jobs
About 1.4 million in 2018

Median Pay
$105,590 in 2018

Educational Requirements
Bachelor's degree

Personal Qualities
Creative, analytical, problem-solving skills

Work Setting
Office environment

Future Job Outlook
Projected 21 percent growth through 2028

software development engineers perform both functions and write software code themselves.

If the software does not work as expected, software development engineers revisit the design process. They make changes to improve the software. This revision cycle can repeat several times until the software meets expectations. Once software is released to the public, software development engineers monitor it to make sure that it functions as intended. They also create maintenance updates and fix any bugs that arise.

Cutting edge software development engineers are getting involved with data science and machine learning. Data science is a branch of software engineering that involves creating meaningful information from large amounts of data. Within data science, machine learning is an application of artificial intelligence that enables computer systems to automatically learn and improve from experience, the way a human brain does. Machine learning engineers focus on developing computer programs and algorithms that can access data, analyze it, and use it to learn without human intervention.

A Typical Workday for a Software Development Engineer

Christine Movius is a software development engineer at the *Washington Post*, where she creates work flow management tools for publishers. In her job, Movius works with a team of software developers, programmers, and engineers. Together they design, create, and manage a suite of software tools called Arc Publishing. The software assists the *Post's* newsroom in tracking the status of news stories, receiving status alerts, assigning tasks to users, and pitching stories to different online platforms and printed publications.

When Movius arrives at her office, she checks in with her team via a chat application, and they discuss what each person will be working on that day. She might write code, test new features, or rework existing code to improve how the software

Software development engineers design the computer programs used worldwide. They build operating systems, business applications, mobile and web applications, computer games, and more.

works for users. Other times she reviews code written by other team members, searches for errors, and suggests improvements before it goes into the main code base. There might be a team meeting to troubleshoot a problem or discuss new technology. Movius says:

> I find my work to be incredibly satisfying because it is both intellectually and creatively challenging, and because I am in charge of my success through the tools I build. Each day presents its new set of problems, for which I have to deliver solutions that not only function properly, but are also presented elegantly to the end user. For example, if a user wants to create a notification, it is not enough for the notification form to simply work; it has to look good, it has to work quickly, and it has to be easy to use.[5]

In Mumbai, India, Bhaskar Dhariyal is a machine learning engineer with ION Energy. Currently, he is working on a project to use machine learning to assess the health of batteries. Often it is difficult to estimate a battery's health because several factors affect it, including cell chemistry, how the battery was used, the temperature it was exposed to, the charging rate, and more. In addition, a battery's health reduces over time. Dhariyal says:

> My work consists of reading research papers to implement them, holding discussions with my boss around the problems we are trying to solve. The work could be identifying the best approach to solving problems or programming, which includes data preprocessing to modelling. Usually, the data is collected through sensors, and there is a high probability of error. Also, usually the data is not in a proper format, so data preprocessing consumes a lot of my time.[6]

After reviewing data sets on battery aging, Dhariyal and his team used the data to create a machine learning–based model to assess battery health. They plan to launch a battery intelligence platform that combines battery use data, artificial intelligence, data science, and other technology to predict and improve the life of lithium-ion batteries.

Education and Training

Software development engineers typically have a bachelor's degree in computer science, software engineering, mathematics, or a related field. Some employers prefer software engineers to have a master's degree. In school, students should take classes that focus on building software so they can learn skills they will need in software development engineering jobs. Many students gain hands-on experience building software through an internship with a software company. Those who plan to focus on machine learning should take advanced mathematics courses to prepare them to perform sophisticated algorithms and computations.

Although most software development engineers do not write code themselves, they should have a strong understanding of computer programming and coding languages such as C++ and Java. In many positions, software development engineers work closely with computer programmers and need to understand the coding required for a new project. Students can prepare by taking computer programming classes and mathematics in high school and college. Even after they land a job, software development engineers must keep up-to-date on the latest technologies and computer languages.

Skills and Personality

In addition to having strong computer and coding skills, successful software development engineers are creative and analytical with good problem-solving skills and attention to detail. To design software that users want, software development engineers must analyze user needs and determine how to meet those needs. Often they are called to inject a little creativity into the process. As development moves forward, they must also be able to solve any issues as they arise. Because the development process can be long and intricate, people working in this field should be able to concentrate on several parts of an application at the same time and pay attention to detail.

Software development engineers interact with many people in their job, which makes having good interpersonal skills indispensable. They must be able to work effectively and efficiently with others who participate in the software's design, development, and coding. Communication skills are also important since they need to give clear written and verbal instructions to others working on a software project and to explain how the software works to users.

Working Conditions

Software development engineers often work in teams with others but can work alone at times. Long work hours are common, especially when a deadline approaches. Frequently, they work in of-

Understanding Customer Needs

"I now realise I am not just a developer, I am an engineer whose job is complex. It is not just about delivering on customer needs but questioning them at times. Engineers have to understand what customers really need and think of new ideas to fulfil those needs. Then, as an engineer, I have to bring those ideas to life to generate a great end-user experience."

—Ranjana Sharma, senior software engineer

Ranjana Sharma, "What Does a Typical Day Look Like for a Software Engineer?," Silicon Republic, May 14, 2019. www.siliconrepublic.com.

fice environments. With internet connectivity, they can often work remotely, checking in with teammates online. Movius says:

> Another terrific aspect of software development is that you can do your job virtually anywhere and at any time. All you need is a computer and an internet connection. I typically report to my office, but I can easily work remotely whenever I want. This is not the case with every software development job, but it is becoming increasingly more common. I have one coworker who is 100% remote, and my team lead has an infant, so he can easily adjust his work day around his family life.[7]

Employers and Pay

The majority of software development engineers work for computer system design firms and software publishing companies. Some are employed by companies in other industries, such as finance, insurance, manufacturing, and engineering services.

As of May 2018, the median annual salary for software developers, which includes software development engineers, was $105,590 according to the Bureau of Labor Statistics (BLS). The

> ### Working Together with Clients
>
> "My role is not just to build software for our clients but to also work alongside them. . . . An important aspect of working together is to make sure that we have a shared understanding, so when the engagement ends there's somebody to carry everything forward technically."
>
> —Bebe Peng, senior software engineer at Pivotal New York
>
> Bebe Peng, "A Day in the Life of a Pivotal Engineer," Built to Adapt, March 20, 2018. https://builttoadapt.io.

median average was slightly higher for systems software developers, $110,000, as compared to application software developers, $103,620.

What Is the Future Outlook for Software Development Engineers?

Jobs for software developers, which includes engineers, is projected to grow 21 percent through 2028, according to the BLS. This growth rate is much faster than the average growth rate of 5 percent for all occupations. This growth is projected to be especially strong for application software developers, with expected growth of 26 percent through 2028.

The main driver behind the growth in this occupation is a projected large increase in the demand for computer software. New applications will be needed on smartphones, tablets, and other mobile devices. As the health care and insurance industries move more tasks online, they will need new software to manage health care policy enrollments and administer policies.

Opportunities for systems software developers will also grow as more products emerge that use software. As more consumer electronics devices and other smart products begin to use computer systems, the demand for people to design and develop these systems will also increase.

In addition, as concerns over computer security increase, more companies will invest in security software to protect digital assets.

Those who are up-to-date with the latest development tools and programming languages will have the best opportunities to land a job as a software development engineer.

Find Out More

Association for Computer Machinery

www.acm.org

This organization is dedicated to advancing computing as a science and a profession. Its website features publications, education opportunities, information about conferences, and local chapters. It also has a learning center that has thousands of books, videos, tutorials, and other learning resources.

Computer Science.org

www.computerscience.org

This website features information about a variety of computer science careers, including software engineer. It has information about what software engineers do, how to become a software engineer, and types of careers in software engineering.

Computer Science Online

www.computerscienceonline.org

This website features information about several computer science careers, including software engineering. Students can explore information about different career paths, emerging careers, top employers, and other areas of interest.

IEEE Computer Society

www.computer.org

The IEEE Computer Society is a leading membership organization devoted to computer science and engineering. Its website includes information about international conferences, publications, a digital library, education opportunities, and local chapters.

Nanotechnology Engineer

What Does a Nanotechnology Engineer Do?

What is it like to work with objects and materials that are so small they cannot be seen with the naked eye? Nanoscale is used to measure atoms, particles, and other objects that are so tiny they can only be seen with powerful microscopes. To put it into perspective, a sheet of paper is about 100,000 nanometers thick. Nanoengineering manipulates materials and processes at the nanoscale. It is an exciting, relatively new area of engineering that can be used in manufacturing, robotics, biomedicine, energy, and other fields. It combines knowledge of mechanical engineering, materials science, electrical engineering, chemistry, physics, and biology to improve lives and further technology through advances in this tiny world.

From the smallest circuits in cell phones to living cells, nanotechnology engineers work with microscopic materials. They find ways to alter them or use them to improve products. The work of a nanotechnology engineer can vary greatly from job to job. Some engineers specialize in research and development on living cells and molecules. Through their work, they develop ways to alter cells in humans. They also work on products to improve health, such as food additives. Other researchers may work on strengthening medicine to make it more effective. Some nanotechnology engineers in the medical field are working to develop gadgets that

A Few Facts

Number of Jobs
About 132,500 in 2016

Median Pay
$96,980 in 2018

Educational Requirements
Master's degree or PhD

Personal Qualities
Analytical, strong problem-solving skills, attention to detail

Work Setting
Laboratory or office

Future Job Outlook
Projected 6.4 percent growth through 2026

can fix problems on the molecular level. Those working with bio-systems are creating ways to store tiny amounts of DNA or other biological materials for testing and manipulation.

In other fields, nanotechnology engineers work with existing materials to make them stronger, more durable, and lighter. Engineers working in national defense have used nanofibers to create lightweight, superstrong materials that can withstand the force of bullets and explosive devices. They have also developed microscopic sensors that can detect tiny amounts of dangerous radiation or chemicals. Nanotechnology is also used to increase sunscreen's ability to reflect harmful ultraviolet radiation and protect sunglasses from scratches. Nanotechnology engineers are also developing innovative ways to test for pollutants and contaminants in the air, water, and ground.

Some nanotechnology engineers work to create tiny electrical materials that can be used in a variety of products. Nanoelectronics will create smaller, more efficient chips, cards, and other computer parts. Using nanoelectronics, the size of electronics products decreases, which reduces electronic waste without sacrificing functionality. Nanotechnology engineers design and test the materials they produce to make sure they perform as intended. If necessary, they revise their design until the material operates effectively.

A Typical Workday for a Nanotechnology Engineer

On a typical workday, a nanotechnology engineer will meet with other scientists and engineers to design and conduct experiments to develop nanotechnology materials, components, devices, or systems. They conduct these experiments in a laboratory and test new methods to process, test, or manufacture nanotechnology materials and products. For example, some inspect and measure thin films of carbon nanotubes, polymers, or inorganic coatings using a variety of cutting edge tools. Outside the lab, nanotechnology engineers document their work and create blueprints of nanotech designs. They research new developments in nanotechnology and determine how they can be applied to their work.

Antonio Susanna is an engineer who works with nanotechnology at the Research and Development Lab at Pirelli, an Italian tire manufacturer. At the lab the engineers research and develop materials to make the company's tires more durable and efficient. Susanna explains that advances in nanomaterials have improved the way his company makes tires, which also reduces the amount of harmful waste dumped into the environment. Zinc oxide is an important material used in tires to transform rubber into a solid during the manufacturing process. However, when zinc oxide from dumped tires leaks into the environment, it causes damage to sea life. "Nanomaterials have been able to improve the properties of the ingredients that make up the [tire], reinforcing it and as a consequence reducing the amount of zinc oxide used, while also maintaining the qualities of resistance and stability in the product," says Susanna. "This is the point of the industry: often nanomaterials enable more effective . . . use of materials."[8]

Yunfeng Shi, an associate professor of materials science and engineering at Rensselaer Polytechnic Institute, is working on a new process that could make glass less likely to break. Currently, the glass used in many smartphones is from the oxide glass family, in which atoms of silicon (a chemical element) bond to four oxygen atoms. This type of bonding is very rigid and allows for very little bending or stretching. When a phone is dropped on the hard floor, the glass cracks. Shi and his team are studying silica (a chemical compound consisting of silicon and oxygen) glass, which is made by compressing silica nanoparticles together. Through molecular simulations in the lab, Shi discovered that silica glass can be stretched up to 100 percent without breaking. This ability to stretch occurs when silicon bonds with five oxygen atoms instead of four. "The compression actually changes the material structure," Shi says. The enhanced plasticity allows the glass to withstand more stress without breaking. Next, Shi's team plans to conduct testing on the silica glass in the lab. The potential uses for silica glass extend beyond smartphones. "This glass is actually as stiff as steel. So, if the glass can be toughened sufficiently, it can replace steel," Shi says. "Our holy grail is to make a transparent structural material."[9]

Changing Properties at the Nanoscale

"Nanomaterials and nanotechnology have become incredibly important for the industry in the last twenty years, because through nanoscales and nano dimensions it's possible to manage multiple properties [of a material]. Taking a material at a macro or a nano level affects its optical, electric, thermal and structural properties: on a nano scale, you get a material with completely different properties to when you work with macro material."

—Antonio Susanna, engineer with Pirelli, an Italian tire manufacturer

Quoted in Pirelli, "Small Is Beautiful: The Power of Nanotechnology Told by Those Who Work with It Every Day," 2016. www.pirelli.com.

Education and Training

Because they are required to have extensive scientific training, students who want to work in nanotechnology engineering should earn at least a master's degree in mechanical, computer, biomedical, chemical, materials, or electrical engineering with a concentration in nanotechnology. Many employers require nanotechnology engineers to have a PhD, so a doctoral engineering degree program that focuses on nanoscience research can be very helpful. High school students interested in this career should take classes in mathematics, physics, chemistry, biology, and engineering.

In addition to core science and engineering courses, undergraduate and graduate students study the fundamentals of nanotechnology and examine how engineering works on a nanoscale. They explore quantum mechanics, atomic structures, and the physics of molecules and solids. They study behavior of materials at atomic levels, micro-electrical mechanical devices and systems, and biomedical engineering at the cellular and subcellular levels. Some choose to take courses that focus on photonics (the study of light) and fiber-optic communications to prepare for a branch of nanotechnology that studies the way nanoscale objects interact with light.

Skills and Personality

In addition to strong science and engineering skills, nanotechnology engineers should have solid analytical and problem-solving

Treating Water Contaminants with Nanotechnology

"My research at the moment is on developing nano materials for water treatment, specifically recyclable materials to treat organic water contaminants. . . . One cool thing about my research is that the particles we have developed enable a completely passive (zero chemical or electricity input) water treatment platform for extremely persistent pollutants like naphthenic acids. This happens to be a great fit for an industry like Canada's oil sands, [which has] a massive inventory of water requiring treatment, but where any treatment solution has to have extremely minimal operating costs."

—Tim Leshuk, PhD engineering student at the Waterloo Institute for Nanotechnology in Canada

Quoted in David Wulff, "Graduate Student Interview: Tim Leshuk," Waterloo Institute for Nanotechnology, January 3, 2018. https://uwaterloo.ca.

skills. They regularly review data related to their research and the products they create. Being able to analyze the data effectively and derive insights that will allow them to improve their work and products is essential. As they review the performance of products, they will need to troubleshoot and solve any issues that arise and develop ways to alter their designs to improve product performance.

Attention to detail is another essential skill in this career. Nanotechnology engineers work with the tiniest objects and must be able to focus on their work and study intricate design plans to find the source of specific problems. Additionally, nanotechnology engineers should have solid written and oral communication skills, since they are frequently asked to write detailed reports and share their research and the results of their testing with others within their company and field.

Having a healthy curiosity and inquisitive nature can also be beneficial in this career. Nanotechnology engineers often work on creating new products, materials, or solutions. Being able to investigate how and why objects behave the way they do can help nanotechnology engineers develop a new product or effective solution.

Working Conditions

Nanotechnology engineers generally work in a lab setting with advanced scientific equipment and computer systems. Since most of the work in nanotechnology engineering is with microscopic objects, the lab will typically have a variety of high-tech microscopes that allow engineers to see and manipulate nanoscale objects. These labs may be found in science research facilities, pharmaceutical companies, medical supply and equipment companies, semiconductor companies, and similar organizations.

Nanotechnology engineers often work closely with others as part of a team to tackle a larger project or objective. "Every project is very closely linked to what your colleagues are doing: the work is multi-dimensional, though it always remains within the company,"[10] says Susanna.

Employers and Pay

The average annual salary for nanotechnology engineers was $96,980 in 2018, according to the CareerExplorer website. Salaries ranged from $50,750 to up to $155,650. Nanotechnology engineers work in almost every industry, including the pharmaceutical, biotechnology, medical, electronics, and more. With nanotechnology's potential to create durable, efficient products across many industries, nanotechnology engineers will be in demand across a variety of industries, working on a multitude of projects.

What Is the Future Outlook for Nanotechnology Engineers?

Nanotechnology is a fast-growing area of science that combines chemistry, physics, biology, engineering, medicine, computer science, materials science, physical science, and other fields. Many industries from computer design to medicine are invested in this new technology. Engineers who have education, training, and experience working in nanotechnology will be highly marketable and in demand.

Jobs for nanotechnology engineers are projected to grow 6.4 percent through 2026, according to CareerExplorer. This growth rate is slightly more than the projected average growth rate for all occupations of 5 percent through 2028, according to the Bureau of Labor Statistics.

Find Out More

American Society of Mechanical Engineers (ASME)
www.asme.org

The ASME is a not-for-profit membership organization that enables collaboration, knowledge sharing, career enrichment, and skills development across all engineering disciplines. A search of its website returns numerous articles about nanotechnology and nanotechnology engineering.

IEEE Nanotechnology Council
https://ieeenano.org

The IEEE Nanotechnology Council strives to advance and coordinate work in the field of nanotechnology carried out in scientific, literary, and educational areas. Its website features a nanotechnology blog, newsletter, publications, and more.

Nano: The Magazine for Small Science
https://nano-magazine.com

This online magazine features numerous articles about the latest nanotechnology news, research, businesses, and funding. Students can learn about the latest advances in the field as well as the research labs and companies where the work occurs.

National Nanotechnology Initiative (NNI)
www.nano.gov

The NNI is a US government research and development initiative involving the nanotechnology-related activities of twenty federal department and agency units. Its website features publications, educational resources, local events, and the latest news and press releases about advances in nanotechnology.

Biomedical Engineer

What Does a Biomedical Engineer Do?

Biomedical engineering combines medicine, biological science, and engineering. Biomedical engineers design the latest high-tech equipment, devices, computer systems, and software used to improve health. They engage in cutting edge research to solve clinical problems and improve the way humans live. For example, biomedical engineers are working on ways to improve prosthetic limbs, revolutionize medicine delivery technology, advance tissue and stem cell research, and introduce a variety of new technologies used in health care.

Within the biomedical field, these engineers design instruments, devices, and software used in health care. Some biomedical engineers design electrical circuits or software to run medical equipment. Others create computer simulations to test new medications. They design and build artificial body parts such as hip and knee joints and limbs. Sometimes, biomedical engineers develop the materials used to build the artificial body parts. They also design and build exercise equipment used for rehabilitation and physical therapy.

One example of the cutting edge work biomedical engineers are doing involves using 3-D printers to create human organs and tissues. The number

A Few Facts

Number of Jobs
About 19,800 in 2018

Median Pay
$88,550 in 2018

Educational Requirements
Bachelor's degree

Personal Qualities
Analytical, good problem-solving and communication skills

Work Setting
Laboratory, hospital, or manufacturing plant

Future Job Outlook
Projected 4 percent growth through 2028

Improving Drug Delivery

"We work on the development of methods to deliver drugs and genes under the guidance of medical imaging. We use all forms of medical imaging and interventional techniques to enhance delivery. We also synthesize unique drug carriers. . . . We find that we can greatly enhance drug delivery—50-fold in some cases—and by combining drug therapy and immunotherapy we see the potential to treat, and in some cases cure, difficult cancers."

—Katherine Whittaker Ferrara, distinguished professor of biomedical engineering

Quoted in Lisa Howard, "4 Innovative Women Engineers Tell Their Story," University of California, June 23, 2016. www.universityofcalifornia.edu.

of patients waiting on organ transplant lists is much larger than the number of organs available. As a result, thousands of patients die of organ failure before they receive a lifesaving transplant. To solve this problem, biomedical engineers are working to develop advanced bioprinters, which are 3-D printers that use a special bio ink to print soft biomaterials. They hope that this technology could one day lead to the printing of replacement tissues, organ parts, or even whole organs that could function in the human body.

In 2019 a company called Biolife4D announced that it had developed technology to print human cardiac tissue using a patient's own cardiac cells. This technology could eventually be used to create whole transplantable hearts that each patient's body will easily accept. "What we're working on is literally bioprinting a human heart viable for transplantation out of a patient's own cells, so that we're not only addressing the problem with the lack of [organ] supply, but by bioengineering the heart out of their own cells, we're eliminating the rejection,"[11] says Biolife4D chief executive officer Steven Morris.

A Typical Workday for a Biomedical Engineer

On any given day, biomedical engineers are involved in a variety of tasks. They spend a lot of time researching new materials to be used in products such as artificial limbs and organs. They develop models or computer simulations of human systems to gather data about the body's processes. They conduct research with other scientists and medical professionals on the engineering aspects of biological and chemical processes. Often they report their research findings to other scientists, executives, clinicians, hospital management, engineers, and others through publication in a scientific journal or an oral presentation.

With all of this information, biomedical engineers often work as part of a team to design and develop new medical instruments, equipment, and procedures for health care. They build models and test prototypes. They analyze results and make any adjustments necessary to the product. When a prototype functions, the team members will present it for review, hoping it will meet with their company's approval and be brought to market.

Once a product has been launched, biomedical engineers install, maintain, and repair the equipment. They demonstrate and explain how to use the equipment to clinicians and other medical personnel. When there is an issue with the device, they provide technical support and respond to customer concerns. They also evaluate the safety, effectiveness, and efficiency of biomedical equipment in the field.

Amanda Edwards is a biomedical engineer who worked as a graduate student on developing treatments for patients with damage to the cerebellum, a part of the brain that is involved in motor skills, coordination, and motor learning. When this part of the brain is damaged, patients exhibit poor coordination and move almost as if they are drunk. Edwards says:

> Unfortunately, treatment for these patients is extremely limited. I developed two different intervention techniques to attempt to improve their movements. As a graduate student, I had the opportunity to get to know these patients personally. I tested my non-invasive interventions on patients who came into the lab. I measured how well the interventions helped by having the patients do virtual reality reaching tasks in an exoskeleton robot.[12]

In her testing, Edwards used a Kinarm exoskeleton robot, an upper-extremity robot with arm trays that support the patient's arms while still allowing elbow and shoulder movement. "The robot has motors which allow us to both drive subjects' movements as well as record their own [voluntary] movements and forces,"[13] says Edwards.

Education and Training

Most biomedical engineers earn at least a bachelor's degree in biomedical engineering or a related engineering field such as mechanical or electrical engineering. Some employers may require candidates to hold a master's degree or PhD. Students in a bachelor's degree program for biomedical engineering focus on engineering and biological science courses. They study fluid and solid mechanics, computer programming, circuit design, and biomaterials. They also take classes in sciences such as biology, chemistry, and physiology.

Students in accredited engineering programs study engineering design. Biomedical engineering programs are accredited by the Accreditation Board for Engineering and Technology. Many programs incorporate co-ops or internships, in which students gain hands-on experience with hospitals, medical devices, and pharmaceutical manufacturing companies. As an undergraduate, Edwards spent a summer in a research experience at Case Western Reserve University. There she worked on a project developing a nerve stimulation device to help stroke patients with swallowing. "This opportunity gave me my best introduction to regular 'lab life.' This allowed me . . . to get experience in the neural engineering field and to put my inventive skills to use in a project that had an important patient-centered goal,"[14] she says.

High school students interested in becoming biomedical engineers should take science courses such as biology, chemistry, and physics. They should also take math classes such as algebra, ge-

Biomedical engineers are working on ways to revolutionize medicine delivery technology, advance tissue and stem cell research, and introduce a variety of new technologies used in health care.

ometry, trigonometry, and calculus. If available, engineering, computer science, and mechanical drawing classes are also helpful.

Skills and Personality

In addition to solid science and technical skills, biomedical engineers should be creative and have strong analytical and problem-solving skills. Every day, these professionals use analytical skills to assess the needs of patients and customers and troubleshoot ways to solve problems. They must be creative to come up with innovative designs for new and improved medical devices and procedures.

Biomedical engineers should also have strong math and computer skills. They frequently use calculus and other advanced math and statistics in analysis and design. As part of the research and design process, they frequently use computers to run simulations

Understanding the Human Body

"A biomedical engineer must be multilingual, as every field of science has its own lingo. For example, in my team, we develop medical devices for diagnosis and treatment of sleep apnea. It is not enough to have electronics knowledge and skill; one must also have in-depth knowledge of upper airway physiology, sleep apnea pathophysiology and the characteristics of the signals that are being recorded. That is why I am a firm believer that a biomedical engineer must have experience in the recording and analysis of basic human biological signals—such as those of the muscles, heart, brain and respiration."

—Zahra Moussavi, professor of biomedical engineering at the University of Manitoba in Canada

Zahra Moussavi, "A War Made Me Realize: The World Needs Biomedical Engineers," The Conversation, October 5, 2017. http://theconversation.com.

to gather data, build virtual models, and perform virtual testing of designs.

Good oral and written communication skills are also beneficial for this career. Biomedical engineers often present the results of their research in journals and presentations. Because they often work as part of a team, they must be able to express themselves clearly and communicate with patients, customers, and team members as they develop new solutions.

Working Conditions

Biomedical engineers generally work on a team with other engineers, scientists, and medical professionals. Their work location varies by project. They may work in a lab, at a hospital or medical facility, or in a manufacturing plant. A biomedical engineer who has designed a new insulin pump for patients with diabetes may spend hours in a medical facility to determine whether the device works as planned. If he or she makes changes to improve the pump, the engineer may go to the manufacturing plant to help rework the manufacturing process for the improved design.

Biomedical engineers typically work a full-time schedule. They may be required to work additional hours to meet deadlines and the needs of patients and customers.

Employers and Pay

Many biomedical engineers work for companies that manufacture medical equipment and supplies. Others are employed by research and development organizations and colleges, universities, and professional schools. Some biomedical engineers work for health care organizations or instrument manufacturers.

As of May 2018, the median annual salary for biomedical engineers was $88,550, according to the Bureau of Labor Statistics (BLS). The lowest-paid 10 percent earned less than $51,890, and the highest-paid 10 percent earned more than $144,350.

What Is the Future Outlook for Biomedical Engineers?

Jobs for biomedical engineers are projected to grow 4 percent through 2028, according to the BLS. This growth rate is about the same as the average growth rate for all occupations (5 percent).

The main driver behind the growth in this occupation is the application of advances in technology to medical equipment and devices. For example, smartphone technology and 3-D printing are cutting edge technologies that are now being applied to biomedical research and equipment. Additionally, as the aging population lives longer and remains active, the demand for biomedical devices and procedures such as hip and knee replacements will increase. As more biomedical advances become mainstream, the number of people seeking biomedical solutions to a wide range of injuries and physical disabilities will also increase.

Find Out More

American Institute for Medical and Biological Engineering
https://aimbe.org

The American Institute for Medical and Biological Engineering is a nonprofit organization headquartered in Washington, DC, that represents top professionals in the medical and biological engineering profession. Its website has a student section with information and resources to help students of all levels.

Biomedical Engineering Society (BMES)
www.bmes.org

The BMES is the professional society for biomedical engineering and bioengineering. Its website includes information about membership, publications, educational resources, bulletins, a blog, and more.

IEEE Engineering in Medicine and Biology Society (EMBS)
www.embs.org

The EMBS is the world's largest international society for biomedical engineers. Its website features information about conferences, committees, journals, and member communities.

Institute of Biological Engineering (IBE)
www.ibe.org

The IBE is a professional organization that encourages inquiry and interest in biological engineering. Its website has several resources for students, including academic programs, a chapter list, a career center, newsletters, journals, and more.

Robotics Engineer

What Does a Robotics Engineer Do?

Working in an area of engineering that is on the rise, robotics engineers are the designers responsible for creating robots and robotic systems. Worldwide, robots are an important part of many production and manufacturing companies. Robots and automated machines never tire. They can work twenty-four hours a day in hazardous conditions and are extremely accurate and precise. Robotic systems perform tasks that human workers cannot or do not want to do. Robotics engineers bring these machines to life, maintain and repair them, and design and build new models. Through their work, robotics engineers make jobs easier, more efficient, and safer for human workers, especially in the manufacturing industry.

A robotics engineer spends a lot of time creating designs and plans needed to build robots. At the beginning of the design process, robotics engineers research exactly what the robot will be needed to do and how it will accomplish its tasks. They may also spend time designing the processes needed for the robot to operate properly. Some robotics engineers design the machines that assemble robots.

After the design process is complete, robotics engineers work to build the robot or robotic system. The building process is frequently tedious and

> **A Few Facts**
>
> **Number of Jobs**
> About 132,500 in 2016
>
> **Median Pay**
> $61,614 in 2018
>
> **Educational Requirements**
> Bachelor's degree
>
> **Personal Qualities**
> Strong coding and mechanical skills, communication skills
>
> **Work Setting**
> Laboratory or office
>
> **Future Job Outlook**
> Projected 6.4 percent growth through 2026

Making a Personal Robot

"9:30 am: Time for the daily standup [meeting]. Each person quickly goes over what they did yesterday, what they're doing today, and whether they need any assistance. Today the Personality team is talking about how to make Misty's [a programmable personal robot] eyes be expressive according to the robot's current emotion. The Vision team is working on a new algorithm to get the robot to recognize its charger. The Programmability team is working on the new skill system [for the robot]."

—Areeya Taylor, robotics engineer at Misty Robotics

Areeya Taylor, "A Day in the Life of a Robot Test Engineer," Misty Robotics, June 12, 2018. www.mistyrobotics.com.

time consuming. Robots are highly technical and difficult to build. It is not unusual for a project to take years to complete.

Today robotics engineers typically specialize in three main skill areas. Some focus on computer-aided drafting and design. They design and improve blueprints for robotic systems. They use cutting edge 3-D modeling programs such as AutoCAD or Blender to create designs and schematics. Other robotics engineers specialize in the hands-on construction of robots, as well as the creation of manufacturing tools and processes that will be used during robot construction. These engineers often work with 3-D printing platforms. Still other robotics engineers focus on research and development, coming up with new and improved robotic designs and uses.

A Typical Workday for a Robotics Engineer

On a typical day, robotics engineers design, build, and test robots. They use computer-aided design (CAD) software to create blueprints and schematics for robotic systems. Some robotics engineers incorporate cutting edge technologies such as artificial intelligence and machine learning to improve the performance of their robotic systems. They may develop software and processes that control the robotic system's functions. They may even design the machines and manufacturing systems to build the robot.

Once the design is complete, robotics engineers build a prototype. They test the prototype and analyze its functions. If they uncover any flaws, they make changes to the design to address the problem. Once prototype testing is complete, robotics engineers work on building and testing the individual parts of a robotic system. Then they assemble the entire robot and test the system as a whole. If they find any errors or the robot does not perform as expected, robotics engineers troubleshoot the problem and make the necessary changes to the design. After the robot is put into use, robotics engineers provide technical support for users. They also research and plan for any new features and improvements to add to the next-generation model.

Since 2012, robots have moved inventory in Amazon's fulfillment centers around the world. Using robots to wheel inventory directly to fulfillment center employees for them to select ordered products saves a significant amount of time in the fulfillment process. In 2019 there were more than one hundred thousand robots working in Amazon's fulfillment centers. In 2015 Amazon engineer Dragan Pajevic led the team that developed the company's next-generation robot. The first-generation robots were 12 inches (30.5 cm) tall, but Pajevic's goal was to design the new robot to be smaller, about 9 to 10 inches (22.9 to 25.4 cm) tall. Using a smaller robot would open up more space in Amazon's fulfillment centers for inventory, allowing the company to use smaller buildings, ship items faster, and lower prices for customers. Pajevic's team designed a robot that was only 7.75 inches tall (19.7 cm) and contained 50 percent fewer components than the earlier model. Called Hercules or H Drive, the new robot can lift 1,250 pounds (567 kg), which is 500 pounds (227 kg) more than the original model. The robot's frame allows its parts to sit closer together, which makes it more compact. The simpler design makes the robot easier to build and maintain. The smaller unit is also better at navigating a fulfillment center. With a smaller, more agile robot, Amazon fulfillment centers are more efficient, and customers are able to get their purchases faster. "Complicated things are not necessarily better than simple things," Pajevic says. "And sometimes it's more difficult to design things to be simple versus complex."[15]

Education and Training

Robotics engineers must have a bachelor's degree in engineering. Because robotics encompasses several engineering disciplines, engineers interested in pursuing a career in robotics will typically earn a degree in electrical, manufacturing, industrial, electronics, or mechanical engineering. Top-notch engineering programs are accredited by the Accreditation Board for Engineering and Technology. Students should also take courses in robotics such as pneumatics and hydraulics, numerically controlled systems, CAD systems, integrated systems, logic, and microprocessors. Some colleges offer engineering programs specifically designed for those interested in robotics. In addition to engineering courses, students should take classes in electronics, chemistry, physics, and mathematics. Classes in software engineering and coding are also beneficial.

Many engineering programs will give students the opportunity to work as an intern at a robotics company to gain hands-on experience. Mikell Taylor, a robotics engineer and senior director of sales operations and customer support with Veo Robotics, says:

> In the days before Robotics Engineering degrees were available, I found my way into the industry by pursuing a degree in Electrical Engineering. I strategically worked my way towards a robotics specialization by choosing robotics projects for classes, getting internships that introduced me to microcontrollers, sensors, and controls, taking extra courses in mechanical and software engineering, and signing up to work with professors who did robotics research.[16]

Using Robots to Automate Genetic Lab Processes

"Robotics Engineers on the [research and development] team are responsible for creating the methods that the [genetic testing] laboratory uses. We write long and complex methods on our liquid-handling robots in order to automate the DNA sequencing process for the lab. It is also our job to predict what Color's [a genetic testing company] future needs will be and build the capabilities required to carry out the company vision. This role is reliant on technical know-how, but it also heavily emphasizes creativity and not being afraid to get out of your comfort zone."

—Tony, robotics engineer

Ben Kobren, "Life at Color—Meet Tony (Robotics Engineer)," Color blog, July 8, 2019. https://blog.color.com.

For upper-level positions, employers may require robotics engineers to have a master's or doctoral degree. As robotics becomes more complex, having a higher-level degree could be essential. Because the industry is constantly developing, robotics engineers will have to stay up-to-date on the latest industry and technical news. Often they accomplish this by attending conferences, seminars, and training sessions.

High school students interested in robotics engineering should take advanced physics and mathematics classes. Classes in computer programming are also beneficial. Students may also choose to join school science or engineering clubs to gain additional experience.

Skills and Personality

In addition to having an engineering degree, robotics engineers should have strong math skills and coding experience. Often robotics engineers are responsible for designing and developing the software applications used to run robots. To build robotic systems, they should also have practical skills in mechanics and electronics. They also need to be problem solvers with a natural curiosity and attention to detail.

Because they frequently work on a team, robotics engineers should have good interpersonal and communication skills. "In the professional world, no *one person* builds a robot," says Taylor. "Teams build robots—teams of people who each specialize in one area, like software development or mechanism design,"[17] Robotics engineers should be able to give direction and discuss updates with other team members. They also need to be able to explain how the robot works to users and listen to any problems or concerns that users have.

Taylor also recommends that robotics engineers develop a specialty in a particular area to make themselves more valuable to prospective employers. She explains:

> Prospective employers often struggle when reviewing multidisciplinary résumés because it's not always obvious how a candidate with multidisciplinary skills aligns with open positions in hardware or software engineering. So if you want to pursue a career at a robotics company, you can increase your chances of success by developing a specialty (through research, side projects, or electives) and making it easier for your employer to identify how you'll fit into a larger team.[18]

Working Conditions

Robotics engineers work in lab and office environments. When in the lab, they work on small components of a robotic system and conduct research. In the office, they work on design plans and write papers. While robotics engineers typically work full time, they may put in extra hours to meet a deadline. Occasional travel may be necessary to visit client sites or manufacturing plants where the robotic system is put into service.

Employers and Pay

Robotics engineers may work in a variety of industries, including the agricultural, automotive, medical, manufacturing, and military industries. Some might develop and innovate the robots used on vehicle assembly lines in the automotive industry. Other robotics engineers might work for companies in food packaging, appliance manufacturing, electronics, and more. Anywhere robots are used to increase efficiency and improve safety, robotics engineers can be found. The average annual salary for a robotics engineer was $61,614 as of 2018, according to the CareerExplorer website. Salaries in this career generally ranged from $52,302 to $98,118.

What Is the Future Outlook for Robotics Engineers?

Currently there are about 132,500 robotics engineers in the United States. According to CareerExplorer, the demand for robotics engineers is projected to grow by 6.4 percent through 2026. As the number of smart devices and computer-driven machines grows worldwide, the need for automation and robotics will also grow. Technology analyists predict that as the capabilities of robots increase, the use of robots and robotic systems in business and homes will skyrocket in the coming years. "There are opportunities in the areas of home automation, toys, home-help for those with medical issues, agriculture and environmental monitoring,"[19] says Professor Ronan Farrell, a senior lecturer in electronic engineering at Maynooth University in Ireland. Robotics engineers who have advanced degrees and coding skills will have the best opportunities to land a job in robotics engineering.

Find Out More

IEEE Robotics & Automation Society
www.ieee-ras.org

The IEEE Robotics & Automation Society works to advance the study and practice of robotics and automation engineering. Its website features information about membership, conferences, publications, and educational opportunities.

International Federation of Robotics

https://ifr.org

The International Federation of Robotics represents robotics around the world, including people in industry, research and development, and industry associations. Its website has links to the latest news on robotics, robot history, position papers, case studies, and more.

National Robotics Education Foundation

www.the-nref.org

The National Robotics Education Foundation supports robotics students, educators, and professionals. Its website has links to the latest robotics news, product news, and educational resources.

Robotic Industries Association

www.robotics.org

The Robotic Industries Association works to promote, educate, and advance robotics and related automated technologies. Its website contains numerous resources to help people get started or advance their robotics education. There is also a listing of safety standards, webinars, upcoming events, and integrator certification training.

Industrial Engineer

What Does an Industrial Engineer Do?

Industrial engineers (also known as materials and process engineers) are an important part of producing and delivering products and services. Their main job is to streamline the production process from start to finish and eliminate wastefulness along the way. Industrial engineers look at the overall production picture and design systems that efficiently integrate workers, machines, materials, energy, and information to create a product or provide a service.

Industrial engineers balance many factors as they strive to deliver a product or service most efficiently. They consider the number of workers required, the technology needed, the steps workers must take, worker safety, environmental concerns, and cost. The end goal is to design a production process that reduces the materials, time, or labor needed while still providing a product or service that has no errors and satisfies customers.

Some industrial engineers are called manufacturing engineers. They concentrate specifically on the automated portions of the manufacturing processes. They design manufacturing systems that most efficiently use computer networks, robots, and materials. Others focus on supply chain management that enables companies

A Few Facts

Number of Jobs
About 284,600 in 2018

Median Pay
$87,040 in 2018

Educational Requirements
Bachelor's degree

Personal Qualities
Strong math, problem-solving, and critical-thinking skills

Work Setting
Office environment and manufacturing plants

Future Job Outlook
Projected 8 percent growth through 2028

to minimize the cost of storing inventory. Industrial engineers also conduct quality assurance activities to make sure the product or service is being provided as intended and customers are happy. They also work in project management to control costs and maximize efficiencies.

Because of their versatile skills, industrial engineers are valuable members of the team for many companies and organizations. They work in industries from manufacturing to health care systems to business administration. For example, they design systems for manufacturers to move heavy materials within a plant. They devise the most efficient systems to deliver goods from a company to customers, which may involve finding the best locations for manufacturing and processing plants. They design systems that pay workers and evaluate workers' job performance.

A Typical Workday for an Industrial Engineer

To design systems that reduce waste and improve performance, industrial engineers start by studying product requirements carefully. They review production schedules, engineering specifications, existing process flows, and other data to understand the methods and activities used in manufacturing and services. They use mathematical models to analyze processes and help them determine how to manufacture parts or products or deliver services efficiently. They also meet with clients and company personnel to discuss product requirements. They talk to vendors about the purchasing process and to manufacturing plant personnel about manufacturing capabilities.

Industrial engineers also work on control systems and procedures. Once they have designed an efficient production system, they develop and enact quality control procedures that are intended to identify and resolve production problems and minimize costs. They design control systems to make sure products meet required quality standards.

Becca Lasky is an industrial engineer for Bombardier Transportation, a manufacturer of planes and trains. She works in the

Finding Solutions on the Factory Floor

"Most of our work must be done on a computer . . . , but a third of the work day is spent out on the factory floor. We wear our jeans, steel-toed boots, and safety goggles, and blend in with the machines as we study them and try to find solutions to either fix them or improve them. Everyone has their own style of how they prefer to work, but at the end of the day, you have to keep in mind you're working with hundreds of other employees, both domestically and internationally, and being flexible to their work dynamics is important."

—Sofiya Kukharenko, industrial and systems engineer at SKF USA

Quoted in Nivaasya Ramachandran, "Meet Sofiya, Industrial and Systems Engineer," Gladeo, 2019. https://gladeo.org.

---

company's Rail Control Solutions group, where she develops solutions such as integrated control systems that manage, control, and supervise train movement. Lasky is also working on automated people movers in airports and positive train control systems, which are designed to automatically stop trains from colliding or derailing because of excessive speed. "My work changes every single day. I am involved in a lot of early design work, reviewing, analyzing, and providing input to those designs. After the designs are implemented, I do an analysis to assure they are working as intended," she says. Lasky enjoys the variety she encounters every day in her job. "What I love most about my job is that it is never static. It keeps me interested and motivated. And I get to work with different types of people, such as designers, dispatchers, and train operators. There are new challenges all the time. I am never bored!"[20] she says.

Laura Silvoy works as a health care systems engineer for Array Advisers, a leader in health care facility design, consulting, and technology. Silvoy completed a transportation study for a medical center that struggles to move patients from inpatient units to diagnostic testing areas efficiently. Instead of using the center's transportation department, the nursing units transported patients themselves. Some departments hired their own transport teams.

Industrial engineers conduct quality assurance activities to make sure a product or service is being provided as intended. They also work in project management to control costs and maximize efficiencies.

To study the problem, Silvoy and her team collected data and used simulation modeling. Silvoy says:

> We took scaled floor plans and traced out every single path a patient can take to get from point A to point B. We then used that model to show what would happen if we took all the patient transportation resources and put them into one group. We found that we could bring all the transporters together and actually allow some of the staff that were not hired exclusively for transporting patients to go back to providing direct patient care. We showed the client that with centralized staff, 98% of the time transport would be done by transport staff and not by a nurse or someone who should be giving care.[21]

Education and Training

Entry-level industrial engineers must have a bachelor's degree, usually in industrial engineering. Programs in industrial engineering are accredited by the Accreditation Board for Engineering and Technology (ABET). Some industrial engineers have degrees in other engineering fields, such as mechanical, electrical, or manufacturing engineering. High school students interested in pursuing a career in industrial engineering should take classes in mathematics, including algebra, trigonometry, and calculus. They should also take classes in computer science, chemistry, and physics to prepare for a college engineering program that will include a broad base of science classes.

In college, industrial engineering students attend classroom lectures and hands-on laboratory sessions. They take classes in statistics, production systems, planning, and manufacturing systems design. Some colleges offer co-op experiences in which students can gain hands-on experience by completing an engineering internship with a company. Laskey participated in a co-op program while she was an undergraduate engineering student at the University of Pittsburgh. In the program, she rotated between school and full-time work assignments in industrial engineering. "I worked full-time for a term with Bombardier, and then returned to school for full time study for a term, through three different rotations. I ended up working for about a year while in college,"[22] she says. After graduating, industrial engineers stay current by attending training classes and seminars.

Industrial engineers who wish to advance to higher-level positions should earn a professional engineering (PE) license. Someone with a PE license can oversee the work of other engineers, approve designs, sign off on projects, and provide services directly to the public. To earn a PE license, candidates must have earned a bachelor's degree from an ABET-accredited engineering program, pass the fundamentals of engineering exam, pass the PE exam, and have at least four years of work experience. Each state has its own licensing process and may require engineers to complete continuing education courses to keep their licenses up-to-date.

Walking the Production Line

"I really like to walk the production line and find things that can be improved. This is done through personal observation and by talking with the associates working on the line. I get to improve things such as procedure, ergonomics [study of worker efficiency], and waste. Not only do these improvements help the company, but they also help the operator."

—Mike Heatwole, industrial engineer with Mars Chocolate North America

Quoted in Kansas State University, "Beyond K-State: Meet Mike Heatwole, '12 BSMS, of Mars Chocolate North America (M&M)," 2016. www.imse.ksu.edu.

Skills and Personality

Industrial engineers must have strong math skills. They regularly use advanced mathematics, including calculus and trigonometry, as they analyze, design, and troubleshoot systems and processes.

In addition to math and technical skills, successful industrial engineers are creative, with strong critical-thinking and problem-solving skills. These engineering professionals are required to use creativity to design production processes that solve problems, reduce waste, and improve efficiencies. Designing new systems and processes requires engineers to use logic and reasoning to analyze the strengths and weaknesses of proposed solutions.

Industrial engineers often work in teams and should have strong oral and written communication skills. They often meet with customers, vendors, and production staff and must be able to listen and communicate well. They must be able to clearly explain processes to production staff and technicians, both orally and in writing. They also prepare written documentation for other engineers and production team members. The documentation must be presented clearly so that it can be easily understood by others.

Working Conditions

Industrial engineers work both in offices and in manufacturing plants or other settings where they are trying to improve processes. For example, in a toy manufacturing plant, they may spend time observing workers as they assemble toy parts in a factory. Later they may return to an office, where they analyze data they have collected. Industrial engineers may be required to travel to various locations to observe processes and make observations in different work settings. While most industrial engineers work full time, the hours may vary based on the projects and industries their work involves.

Employers and Pay

In 2018 there were about 284,600 jobs for industrial engineers, according to the Bureau of Labor Statistics (BLS). Industrial engineers typically worked for employers in transportation equipment manufacturing (18 percent); computer and electronic product manufacturing (13 percent); professional, scientific, and technical services (12 percent); machinery manufacturing (8 percent); and fabricated metal product manufacturing (5 percent).

As of May 2018 the median annual wage for industrial engineers was $87,040, according to the BLS. The lowest-paid 10 percent of industrial engineers earned less than $56,470, while the highest-paid 10 percent earned more than $132,340.

What Is the Future Outlook for Industrial Engineers?

Job opportunities for industrial engineers are projected to grow about 8 percent through 2028, according to the BLS. This rate is higher than the projected average growth rate for all occupations (5 percent). Because their main focus is reducing costs and wastes and improving efficiencies, industrial engineers are in demand for projects in a wide range of industries. For example, as the health care industry grows and changes are made in how health care is delivered, industrial engineers will be needed to design efficient systems and control procedures.

In addition, as new technologies to automate production emerge, industrial engineers will be needed to incorporate automation technologies in the production process for many industries.

Industrial engineers with knowledge and experience in manufacturing engineering will have the best opportunities to land jobs in the field.

Find Out More

Institute of Industrial & Systems Engineers (IISE)
www.iise.org

The IISE is the world's largest professional society dedicated to the support of the industrial engineering profession and individuals involved with improving quality and productivity. Its website includes a Student Center, which has information about the profession, scholarships, competitions, student membership, and other resources.

National Society of Professional Engineers
www.nspe.org

The National Society of Professional Engineers is a national organization committed to addressing the professional concerns of licensed professional engineers across all disciplines. Its website has information about continuing education, industry news, and licensing.

Society of Manufacturing Engineers (SME)
www.sme.org

The SME is a nonprofit association of professionals, educators, and students committed to promoting and supporting the manufacturing industry. Its website offers information about events and trade shows, training courses, manufacturing news, and more.

Technology Student Association (TSA)
https://tsaweb.org

The TSA is a national organization of students engaged in science, technology, engineering, and mathematics. Its website includes information about competitive events, conferences and programs, learning and career resources, and more.

Environmental Engineering Technician

What Does an Environmental Engineering Technician Do?

Environmental engineering technicians work with environmental engineers to develop solutions to environmental problems. For example, many communities need solutions to improve water and air quality. Industries need help treating and containing hazardous waste. Wildlife and ecosystems need protection from human activities. Communities are searching for ways to produce energy in cleaner, more environmentally friendly ways.

Environmental engineers and technicians use the principles of engineering and knowledge of soil science, biology, and chemistry to devise environmental solutions. They try to improve the processes of recycling, waste disposal, public health, and water and air pollution control. They also work on worldwide issues such as making drinking water safe, climate change, and environmental sustainability.

Environmental engineering technicians often work in teams with other technicians and engineers. Working from plans created by environmental engineers, technicians perform a variety of tasks. They collect air, water, soil, and other samples and perform tests on these samples. Technicians

> **A Few Facts**
>
> **Number of Jobs**
> About 17,900 in 2018
>
> **Median Pay**
> $50,560 in 2018
>
> **Educational Requirements**
> Associate's degree
>
> **Personal Qualities**
> Critical-thinking and analytical skills, communication skills
>
> **Work Setting**
> Office environment and outdoor sites
>
> **Future Job Outlook**
> Projected 9 percent growth through 2028

Monitoring Environmental Conditions at a Mine

"I head out to the [open-pit mining] operation's tailings and reclamation ponds [which contain materials from mining activities, water runoff, and rainwater]. These contain the leftover materials from processing borates, and contain arsenic and other substances which can be harmful to wildlife. I usually make two or three trips to the ponds each day to check equipment, monitor key environmental targets such as air quality and water levels, and record wildlife. We have around 700 to 1,000 birds visiting our ponds every month, and during migration season we can spend up to six hours a day recording and tracking birds."

—Larry Fealy, senior environmental technician at Rio Tinto's Boron Operations, an open-pit mine in California

Larry Fealy, "A Day in the Life: Larry Fealy," RioTinto, September 30, 2016. www.riotinto.com.

---

also spend time working with engineers to develop plans and create devices to reduce pollution. Some technicians travel to inspect various facilities to make sure they are complying with environmental regulations.

Joe Proulx is an applied engineering technology instructor at Northcentral Technical College in Wausau, Wisconsin. He believes that working as an environmental engineering technician can be very rewarding. "You're doing the right thing to protect the environment—that in itself can be a great reward," he says. "If you're able to come up with a plan or make an improvement [to the environment], it's rewarding to know what you're doing for the local community."[23]

A Typical Workday for an Environmental Engineering Technician

On any given day, an environmental engineering technician may spend time in the field or in a laboratory. In the field, technicians collect samples of air, groundwater, soil, and more for pollution surveys. They take photographs of testing areas and potential pollution sources. When hazardous materials such as lead and asbestos are discovered, technicians make arrangements to have these materials disposed of safely. Technicians also set up and operate equipment in the field to prevent or clean up environmental pollution. They conduct inspections of different locations and facilities to assess compliance with environmental regulations and laws.

In the laboratory, environmental engineering technicians test and analyze the samples that they collected in the field for pollution surveys. They record their observations, test results, and photographs and maintain project records and computer files. They review work plans and schedule future testing and other activities. When a pollution problem is discovered, technicians meet with other engineers and technicians to create plans to find and reduce the sources of the pollution. They decontaminate and test field equipment used to test pollutants in the soil, air, or water. Technicians are also responsible for making sure the laboratory is fully stocked. They order materials and equipment, research vendors and suppliers, and gather product information for the lab.

As a lifelong nature lover and camping enthusiast, Ari Cheremeteff chose a career as an environmental engineering technician. She works for Lu Engineers in Rochester, New York, and focuses on revitalizing urban areas that have fallen into disrepair. "I want to see Rochester use a lot of the spaces that are unused because of environmental issues," she says. "To bring function to the city, but while keeping humans from being exposed to toxins."[24] In her role Cheremeteff oversees the environmental areas of the city's redevelopment projects. For example, at a contaminated building site, she coordinates and works with contractors to inject bioremediation material into the soil. The material causes naturally occurring bacteria to grow and then eat petroleum and other toxins in the contaminated soil. Cheremeteff calculates and maps the optimal injection points. She also performs testing to make sure the remediation efforts are working. "Each site is very different, so finding the right plan or approach takes a lot of research and collaboration with people above me," says Cheremeteff. "It's a fun challenge. That's part of the allure for me."[25]

For Arianne, an environmental engineering technician with earth services company Summit, every day brings a new challenge. She routinely tackles complex technical problems to remediate contaminated sites. Every site is different and requires its own specific solution. Arianne identifies each piece needed for the solution, communicates the plan to the remediation team, gathers the relevant safety and technical guidelines, and works on the remediation efforts herself. About 60 percent of her time is spent working on projects outside. She describes an average day:

> I usually start in the office—preparing a site-specific plan and compiling the safety and technical guidance needed. Then I head to the field. Once at the site I review the conditions and safety documentation and we have our daily safety meeting. Then the work begins—monitoring soil and water, taking detailed notes, observations and recording the instrument readings. Throughout the day I continually assess the conditions and adjust my work to accommodate the environment, technical or administrative issues. Then, it's back to the office to prepare a report of the site conditions and activities and analyze the data from the site against regulatory guidelines.[26]

Education and Training

To be considered for most positions, environmental engineering technicians should have at least an associate's degree in environmental engineering technology or a related field. To earn this degree, students can attend a two-year program at a vocation-technical school or community college. Both types of programs include classes in mathematics, chemistry, hazardous waste management, and environmental assessment. Programs at a community college generally also include liberal arts classes. When choosing a program, students should look

Responsibilities on the Job

"While in the field, my duties include assisting senior field staff with groundwater and surface water monitoring [around Kingston, Ontario, Canada], installation and maintenance of monitoring wells, conducting small scale Designated Substance Surveys (DSS), and conducting site observations on various long-term projects. These projects may be local, or as far afield as the Greater Toronto Area or even Sudbury. My office duties include treatment and validation of our field data, AutoCAD and other computer assisted data presentation programs, sample handling, and assisting project managers with report redaction and Environmental Site Assessments (ESAs)."

—Zachary Leger, junior environmental technologist at Malroz Engineering

Quoted in St. Lawrence College, "Environmental Technician Alumni Profiles." www.stlawrencecollege.ca.

for an engineering technology program that is accredited by Accreditation Board for Engineering and Technology (ABET). To prepare for a career as an environmental engineering technician, high school students should take classes in mathematics as well as natural sciences such as biology, chemistry, and physics.

Some environmental engineering technicians have a bachelor's degree in biology or chemistry. In addition, some positions require technicians to have training in working with hazardous materials in accordance with Occupational Safety and Health Administration standards. Environmental engineering technicians generally start as trainees in an entry-level position. They learn the job under the supervision of an environmental engineer or an experienced technician. As they become more experienced, technicians take on more responsibility and work with less supervision. Some advance to become senior environmental technicians or lead technicians and supervise other team members on projects. Technicians who earn a bachelor's degree can advance to become environmental engineers.

Skills and Personality

Successful environmental engineering technicians are good problem solvers and have strong critical-thinking skills, which help them evaluate various solutions to problems. They help environmental engineers identify environmental problems and propose solutions. "The biggest asset I can bring to my job is analytical thinking," says Arianne. "The complexity of the site can be challenging, and I need to use my technical knowledge and understand the relevant regulatory requirements to find the right solution. I'm also balancing stakeholder expectations and collaborating with colleagues, industry experts and clients."[27]

Environmental engineering technicians are the eyes and ears of the team. Therefore, they need to have a keen eye and strong observational skills as they travel to different locations, assess the situation, recognize problems, and quickly inform environmental engineers about what they have observed.

Because they often work in a team with other engineers and technicians, environmental engineering technicians need to have strong communication and interpersonal skills. They should be able to listen to others and clearly communicate their findings, orally and in writing. Strong reading skills are also a benefit for this career, since environmental engineering technicians also spend time reading complex legal and technical documents to make sure regulatory requirements are being met.

Working Conditions

Environmental engineering technicians typically work in a laboratory under the direction of engineers and as part of a team with other engineers and technicians. Sometimes, they travel to outdoor sites that can be in remote locations.

When helping with environmental cleanup, environmental engineering technicians can be exposed to hazardous waste, chemicals, toxic materials, and equipment hazards. In these situations, they wear protective suits and even respirators to guard against injury and contamination.

Most environmental engineering technicians work full time with regular business hours. Sometimes they work irregular and extra hours when monitoring operations or dealing with a major environmental threat.

Employers and Pay

As of May 2018 the median annual wage for environmental engineering technicians was $50,560, according to the Bureau of Labor Statistics (BLS). The highest-paid 10 percent of technicians earned more than $82,500, while the lowest-paid 10 percent earned less than $32,380.

In 2018 many environmental engineering technicians worked for engineering services companies (21 percent) and management, scientific, and technical consulting firms (20 percent), according to the BLS. Other technicians found jobs with government agencies (18 percent), waste management and remediation companies (11 percent), and manufacturing companies (7 percent).

What Is the Future Outlook for Environmental Engineering Technicians?

Job opportunities for environmental engineering technicians are projected to grow 9 percent through 2028, according to the BLS's *Occupational Outlook Handbook*. This growth rate is faster than the projected average growth rate for all occupations (5 percent) over the same period.

As more communities place greater importance on preventing environmental problems, governments and employers will need trained environmental engineers and technicians who are able to incorporate cutting edge environmental technologies in pollution prevention. Additionally, as state and local governments focus more on efficient water use, storm water management, and wastewater treatment in the coming years, the demand for environmental engineering technicians is expected to rise.

Find Out More

American Academy of Environmental Engineers & Scientists
www.aaees.org

The American Academy of Environmental Engineers & Scientists is a not-for-profit organization that provides certification to those who qualify through experience and testing. Its website has information about training, periodicals, student resources, and other resources.

American Society for Engineering Education
www.asee.org

The American Society for Engineering Education promotes innovation, excellence, and access for all levels of education in the engineering field. Its website includes industry news, publications, career information, and information about events and conferences.

National Society of Professional Engineers
www.nspe.org

The National Society of Professional Engineers is a national organization committed to addressing the professional concerns of licensed professional engineers across all disciplines. Its website has information about continuing education, industry news, and licensing.

Technology Student Association (TSA)
https://tsaweb.org

The TSA is a national organization of students engaged in science, technology, engineering, and mathematics. Its website includes information about competitive events, conferences and programs, learning and career resources, and more.

Renewable Energy Engineer

What Does a Renewable Energy Engineer Do?

Much of the world relies on fossil fuels for energy. Fossil fuels such as coal, oil, and natural gas are formed below the surface of the earth from natural matter that decomposes and is buried underground. Over millions of years, the matter is exposed to enormous pressure and heat and eventually forms fossil fuels. However, the world's supply of fossil fuels is continually decreasing. According to some estimates, the world's crude oil supplies could be depleted in the next one hundred years. In addition, the excessive use of fossil fuels continues to have a devastating impact on the earth's environment and climate. The burning of fossil fuels contributes to climate change through the release of carbon dioxide into the atmosphere. Also, the use of fossil fuels releases chemicals and pollutants into the atmosphere, increasing health risks to the world's population.

Around the world, more people are working to develop reliable and efficient renewable energy (or green energy) sources to replace dwindling supplies of fossil fuels. Renewable energy is constantly and naturally replenished. Additionally, using renewable energy has less of an environmental impact than using fossil fuels for energy. Examples

A Few Facts

Number of Jobs
About 312,900 in 2018

Median Pay
$71,446 in 2018

Educational Requirements
Bachelor's degree

Personal Qualities
Strong math and science skills, innovative

Work Setting
Laboratory and outdoor sites

Future Job Outlook
Projected 4 percent growth through 2028*

* Number is for mechanical engineers, which includes renewable energy engineers

> ### Servicing Wind Turbines in Dangerous Conditions
>
> "My primary role is the service and maintenance of our offshore wind turbines at Walney Offshore Windfarm. This includes inspections, scheduled maintenance and troubleshooting any small faults that come up and fixing them. I lead a small team and together we perform a range of manual tasks to ensure that the wind turbines are working reliably and at maximum efficiency. Because we're out at sea and sometimes in challenging weather conditions, ensuring it's done safely is probably the biggest part of my job."
>
> —Conor Lewis, engineer on an offshore wind farm
>
> Quoted in Tomorrow's Engineers, "A Big Fan of Wind Energy," 2019. www.tomorrowsengineers.org.

of renewable energy sources include solar power, wind energy, hydroelectric energy, and geothermal energy.

Renewable energy engineers research and design renewable energy systems such as hydroelectric dams, solar photovoltaic cells, wind turbines, and geothermal plants. They design the systems to carry electricity generated from wind farms, hydroelectric dams, and solar fields to power plants, where it can then be distributed to users. Renewable energy engineers devise ways to turn corn into ethanol or landfill gas into fuel. They investigate ways to improve existing energy projects to make them more efficient and environmentally safe.

A Typical Workday for a Renewable Energy Engineer

Robert Mattholie is an engineer who works for Renewable Energy Systems in southwestern England. In his role he has responsibility over several operational wind and solar energy projects. Each day, he tackles a wide range of responsibilities, from organizing and managing maintenance contractors to responding to failures at any of the power plants. He also identifies and implements opportunities to maximize energy generation at the wind and solar farms and manages all aspects of the renewable energy assets. Mattholie says:

The nature of working on an operational project means that you're invariably working in a fast-paced environment. The range of projects with different technologies means you're exposed to wide-ranging problems (technical, commercial and social) which definitely keeps you on your toes! I also enjoy the opportunity to work with a wide range of contractors and technical specialists, which involves a good mixture of site and office work.[28]

Mattholie says that the unpredictable nature of the job can be one of its most challenging aspects, particularly balancing daily job tasks with unplanned and often urgent events. "You might start the day with various plans but often these can be thrown completely out the window by 9am if something crops up on site,"[29] he says.

Laura O'Keefe is a renewable energy engineer and Innovation Fellow at Lancaster University in England, working with businesses on renewable energy initiatives. She is currently working with River Power Pod, a developer of small, transportable water turbines that can be used to give remote communities the ability to create renewable energy from local streams. O'Keefe says:

We're currently testing their newly developed and patented river turbines. At the moment, I'm part of a team that's measuring and recording the power output of streams against their velocity—without ever losing sight of the environmental impact. Consequently, my days are spent using flow metres in streams and rivers to assess the potential effectiveness of turbines. Historically, this type of energy has been harnessed at schemes based around weirs [low dams], but in the future there will be the potential for communities to generate renewable energy from rivers without the installation of heavy infrastructure.[30]

If successful, this research could be life changing for millions of people around the world. According to O'Keefe, about 1.3 billion people worldwide do not have direct access to electricity.

About 170 million of those live within a mile of a river or stream and could benefit from the electricity-generating small water turbines. "I've always been passionate about renewable energy and this innovation has the potential to make a massive difference globally," says O'Keefe. "Professionally and personally, it's an incredibly satisfying project to be a part of."[31]

Education and Training

Renewable energy engineers must have a bachelor's degree in engineering, typically from a program accredited by the Accreditation Board for Engineering and Technology. Some students earn a degree in renewable energy or energy engineering. Because renewable energy encompasses several engineering disciplines, others earn a degree in a related engineering field, such as mechanical, industrial, electrical, or chemical engineering. "Mechanical, chemical, industrial and electrical engineers work in the alternative energy field," says Vita Como, senior director of professional development at the University of Houston's Cullen College of Engineering Career Center. "Any of the main engineering disciplines can have an alternative energy component."[32]

Students in renewable energy programs learn how to design, install, and repair green energy systems. They take classes in electrochemistry, energy management, fluid mechanics, solid-state devices, statistics, and thermodynamics. In their course work, students study both traditional and renewable energy sources. They also study the function and design of electrical machines. In addition to engineering courses, students should take classes in electronics, chemistry, physics, and mathematics. Many engineering programs will give students the opportunity to work as an intern at an engineering company, where they will gain hands-on experience with real-life engineering projects.

For upper-level positions, employers may require renewable energy engineers to have a master's or doctoral degree. Because the industry is constantly developing, renewable energy engineers will have to stay up-to-date on the latest industry and tech-

> ### Designing a Solar Energy Solution
>
> "Earlier this year, [the company] Solarcentury launched Sunstation. It's a new solar system that is completely different to others because it sits in the roof rather than on it. Developing this new product has been my main focus since I joined the company. I was responsible for designing the perimeter kit which connects the solar panels to the roofing tiles. This consists of seven different components which come together to form an excellent weather and fire barrier. These components needed to be easy to install, low cost and compatible with a range of roof tiles, all making for a complex design challenge."
>
> —Hannah Eastwell, product development engineer at Solarcentury, a solar power company
>
> Quoted in Solarcentury, "Day in the Life: Hannah Eastwell—Product Development Engineer at Solarcentury," June 19, 2017. www.solarcentury.com.

nical news. Often they accomplish this by attending conferences, seminars, and training sessions.

High school students interested in renewable energy engineering should take advanced science and mathematics classes, including biology, chemistry, physics, and calculus. Classes in computer programming are also beneficial. Students may also choose to join school science or engineering clubs to gain additional experience.

Skills and Personality

In addition to having an engineering degree, renewable energy engineers should have strong math and science skills. These skills, along with the ability to problem solve, enable them to take scientific principles and apply them in the design and building of useful renewable energy systems and devices. In addition, creativity is essential to designing and developing complex renewable energy systems and devices, while having mechanical skills is helpful during the building process. "You must have a sound technical and engineering background, the ability to think outside the box and be prepared to innovate to overcome challenges,"[33] says Mattholie.

Renewable energy engineers research and design renewable energy systems such as wind turbines. They design the systems to efficiently carry the energy that is generated to power plants.

Because they frequently work on a team with other engineers and technicians, renewable energy engineers should have good interpersonal and communication skills. They must be able to listen to other team members and analyze different suggestions to finish the current task. On the job, renewable energy engineers also interact and communicate with equipment installers and construction workers.

Working Conditions

Renewable energy engineers typically work in a laboratory as part of a team with other engineers and technicians. They also travel to outdoor work sites, which can be in remote locations. They can work on a variety of projects, from large-scale municipal renewable energy projects to residential home installations.

Most renewable energy engineers work full-time with regular business hours. Sometimes, they work irregular and extra hours when monitoring operations, dealing with unexpected problems, or meeting an approaching deadline.

Employers and Pay

According to PayScale, the average salary for a renewable energy engineer was $71,446 in 2018. Entry-level renewable energy engineers can expect to earn less, while those who have years of experience in the field can earn more than $92,000.

In 2018 there were about 312,900 jobs for mechanical engineers, which includes renewable energy engineers, according to the Bureau of Labor Statistics (BLS).

What Is the Future Outlook for Renewable Energy Engineers?

Job opportunities for all engineers are projected to grow about 4 percent through 2028, according to the BLS. This rate is about the same as the projected average growth rate (5 percent) for all occupations. The demand for renewable energy engineers is one of the drivers of the projected job growth in engineering. The BLS indicates that wind energy is the fastest-growing source of renewable energy. By 2030 wind energy could provide 20 percent of the country's electricity supply. In addition, solar and hydropower sources of energy also have potential to grow in the coming years. As a result, renewable energy engineers will be needed to design, develop, operate, and maintain the growing number of renewable energy facilities.

Additionally, more people are driving alternative vehicles, including cars and trucks that run on electricity, hydrogen, and other alternative fuels. As the demand for vehicles that create less pollution and use renewable energy rises, more renewable energy engineers will be needed to apply their creativity and technical skills to design and develop energy-efficient, high-performance vehicles.

Find Out More

American Academy of Environmental Engineers & Scientists

www.aaees.org

The American Academy of Environmental Engineers & Scientists is a not-for-profit organization that provides certification to those who qualify through experience and testing. Its website has information about training, periodicals, student resources, and other resources.

American Society of Mechanical Engineers (ASME)

www.asme.org

The ASME is a not-for-profit membership organization that enables collaboration, knowledge sharing, career enrichment, and skills development across all engineering disciplines. A search of its website returns numerous articles and resources about renewable energy engineering.

National Society of Professional Engineers

www.nspe.org

The National Society of Professional Engineers is a national organization committed to addressing the professional concerns of licensed professional engineers across all disciplines. Its website has information about continuing education, industry news, and licensing.

Technology Student Association (TSA)

https://tsaweb.org

The TSA is a national organization of students engaged in science, technology, engineering, and mathematics. Its website includes information about competitive events, conferences and programs, learning and career resources, and more.

Modern Structural Engineer

What Does a Structural Engineer Do?

Everywhere people look, they see the work of structural engineers. Specialists in the field of civil engineering, structural engineers are responsible for making sure that the structures people use daily, such as tall buildings and bridges, are safe and stable. Structural engineers make sure bridges can bear heavy loads and roofs do not collapse under the weight of thick snow. They ensure that tall skyscrapers can withstand the strongest gusts of wind. To do this, structural engineers use knowledge of physics, mathematics, and engineering to select construction materials and create designs for structures that can safely hold up under the stresses they encounter, including gravity, storms, and earthquakes.

Structural engineers are at the cutting edge of the design world. They find creative ways to build functional and durable structures that meet the needs of local communities. They stay up-to-date on the latest design and building technologies as well as the latest construction materials. They design all types of structures for clients in both the private and public sectors.

Structural engineers are involved in almost every step of the building process. They assist in the initial de-

A Few Facts

Number of Jobs
About 326,800 in 2018*

Median Pay
$86,640 in 2018

Educational Requirements
Bachelor's degree

Personal Qualities
Analytical, problem-solving, and organizational skills

Work Setting
Office environment and construction sites

Future Job Outlook
Projected 6 percent growth through 2028

* Number is for civil engineers, which includes structural engineers

New Challenges Every Day

"Every day is something different, which creates a dynamic and exciting workplace. By working on various projects, I have been able to use [engineering] programs such as Risa, Safe, Etabs, RetainPro, Autocad and SPColumn, as well as many in-house programs, to assist in the design process. This experience has not only increased my knowledge of the structural engineering concepts, but also reinforced my confidence in my ability."

—Josh Raney, structural engineering intern

Josh Raney, "So, What Does an Internship in the Field of Structural Engineering Really Look Like?," Watry Design, 2019. https://watrydesign.com.

sign of a structure to make sure it is structurally sound and safe for its intended purpose. After the design process, structural engineers take an architect's finished design and calculate the load stresses on the structure—such as gravity, snow, or other external factors—and determine whether the designed structure will be stable under these conditions. They also help architects and builders select the best materials for the structure's purpose to provide the structural integrity the project needs. Sometimes, structural engineers inspect and evaluate existing structures to make sure they meet modern building codes.

A Typical Workday for a Structural Engineer

Every day, structural engineers handle a variety of tasks and responsibilities. They meet with clients, other engineers, and supervisors to review technical project details. They review their state's current standards and regulations to make sure designs are in compliance with all requirements. They research and study different types of building materials to pick the best options for a safe and sustainable structure. They prepare reports and presentations for clients and superiors.

When working on a new project, structural engineers first prepare schematic designs that include basic measurements and layouts. These designs can be used to make sure the plans comply with government requirements. Then the engineers create more detailed design drawings that show building materials and elevations. Finally, they create construction documents that include all the necessary details and measurements required for construction.

In the field, structural engineers inspect job sites and meet with construction workers and architects to review the project's details and progress. They examine a concerning area of a structure to assess damage and determine whether the structure can be repaired or must be entirely replaced.

Katie Symons is a structural engineer who works for Whitbybird, a structural engineering consulting firm based in London. Most of the time, she works in an office preparing designs. Symons says:

> Lots of my structural design work is done by hand, using pencil, paper and calculators as well as sketching paper to come up with structural elements that will work. However, I use computer programs to help speed up the process, and model more complex, or very large structures. Lots of my job is working with the design team for my projects, normally made up of the architect, the mechanical & electrical engineer (who sort out the drainage, water, power, ventilation systems for a building) and any other people who give specialist input.[34]

Education and Training

Structural engineers need to earn a bachelor's degree, ideally in civil engineering or a related field. Some students choose to earn a master's degree with a specialization in structural engineering or similar field, which can improve their chances of competing for higher-level structural engineering positions. Programs in civil engineering include classes in math, statistics, engineering mechanics and systems, and fluid dynamics. Courses are a mix of traditional classroom, laboratory work, and fieldwork.

Some students participate in cooperative programs, or co-ops, in which students spend time working and gaining real-life experience while earning a degree. As a student completing his master of science at California Polytechnic State University, Josh Raney worked as an intern for Watry Design. Raney says:

> On my first day, I immediately began working on an active project. In the first few hours of my internship, I had already completed more "real world" assignments than I had in five years of college. The theoretical and scholastic ideology I had become comfortable with was instantly overshadowed by the physical and material realism that is the workforce. From that point on, I have been exposed to numerous projects in different states, all with various design scenarios.[35]

Structural engineers who wish to advance to higher-level positions should earn a professional engineering (PE) license. Someone with a PE license can oversee the work of other engineers, approve designs, sign off on projects, and provide services directly to the public. To earn a PE license, candidates must earn a bachelor's degree from an engineering program accredited by the Accreditation Board for Engineering and Technology, pass the fundamentals of engineering exam, pass the PE exam, and have at least four years of work experience. Each state has its own licensing process and may require engineers to complete continuing education courses to keep their licenses up-to-date. In addition, some states require structural engineers to earn a separate license by passing the structural engineering (SE) exam. The SE exam tests an engineer's ability to safely design structures while considering risk factors such as high winds and earthquakes.

Skills and Personality

Structural engineers work closely with other professionals to complete a project, from architects to other engineers. As they collaborate, they must be able to communicate their ideas and findings in a clear and understandable way to people with varying

Structural engineers are involved in almost every step of the construction process. They assist in creating the initial design of a structure to make sure it is structurally sound and safe for its intended purpose.

technical backgrounds. They also must be able to explain their work to community leaders and prepare written reports that present technical concepts clearly.

In addition, structural engineers need to be creative and have superior mathematical and technical skills to come up with cutting edge designs that meet client and community needs. They often use the principles of calculus, trigonometry, and other advanced math in their work.

The best structural engineers have strong problem-solving skills and are good project managers. Often they need to identify and analyze complex problems and develop efficient and safe solutions. They should also be proficient in the use of design software and other industry technologies. "I'm always dealing with people from lots of different backgrounds—architects, other engineers, on-site operatives and clients—and I'm always solving problems," says Kate, a structural engineer for Atkins, a global engineering consulting firm. "Sometimes there's lots of problems to solve in the day to keep the construction process moving, other times it's a

> ### Watching the Team Succeed
>
> "I love everything about being a Structural Project Engineer, watching my projects turn from an architect's perspective into a physical concept and knowing it was a team effort. I love doing my own design work and drafting work, yet also the delegation and working with our technicians as a team and having their input and knowledge applied."
>
> —Grace, structural engineer with Rolton Group, an engineering consulting firm
>
> Quoted in Power Engineering International, "A Day in the Life of an Engineer," June 22, 2019. www.powerengineeringint.com.

few bigger design-based problems which can take days or weeks to work through."[36]

Structural engineers balance conflicting objectives, such as customer wants and financial costs and safety concerns. They must be able to make good decisions, often based on their knowledge of industry best practices, technical knowledge, and prior experience. Because they often balance several projects at once, they need good organizational skills to be able to monitor and evaluate progress at a job site, balance time needs, and allocate resources.

Working Conditions

Structural engineers work both indoors and outdoors. When working on designs, structural engineers typically work in an office setting. When a project is in the field, engineers may be required to spend much of their time outside at construction sites so they can monitor the construction progress and address problems that arise at the job site. Sometimes, structural engineers may work in an office in a trailer at a job site. Most structural engineers work full time, with regular business hours. Some structural engineers work more than forty hours per week, especially when a project is under construction. Additional hours may be required to make sure project deadlines are met.

Employers and Pay

Structural engineers work for a variety of entities. Some work for local, state, and federal government agencies to design public structures. Others work in industry on the design and construction of private structures. Structural engineers are also found at consulting firms, working with a variety of clients and projects.

As of May 2018 the median annual wage for civil engineers, which includes structural engineers, was $86,640, according to the Bureau of Labor Statistics (BLS). The highest-paid 10 percent of civil engineers earned more than $142,560, while the lowest-paid 10 percent earned less than $54,780.

What Is the Future Outlook for Structural Engineers?

Job opportunities for civil engineers are projected to grow 6 percent through 2028, according to the BLS. This growth rate is slightly faster than the projected average growth rate for all occupations (5 percent) over the same period.

As infrastructure around the United States deteriorates, civil and structural engineers will be needed to manage projects to rebuild, repair, and upgrade buildings, bridges, dams, and other structures. As the population grows, civil and structural engineers will be needed to design and build new water systems and waste treatment plants and repair aging water systems. Additionally, as more communities invest in renewable-energy projects, civil and structural engineers will be involved in the design of wind farms, solar systems, and similar projects.

Structural engineers who have hands-on work experience, either through previous jobs or by participating in a college co-op program, will have the best opportunities to land a job as a structural engineer. In addition, structural engineers who earn a graduate degree and are licensed will also have an advantage when interviewing for higher-level positions.

Find Out More

American Society for Engineering Education
www.asee.org

The American Society for Engineering Education promotes innovation, excellence, and access for all levels of education in the engineering field. Its website includes industry news, publications, career information, and information about events and conferences.

American Society of Civil Engineers (ASCE)
www.asce.org

The ASCE represents civil engineers worldwide. It provides technical and professional conferences and continuing education, publications, news, and information about current industry issues.

National Society of Professional Engineers
www.nspe.org

The National Society of Professional Engineers is a national organization committed to addressing the professional concerns of licensed professional engineers across all disciplines. Its website has information about continuing education, industry news, and licensing.

Technology Student Association (TSA)
https://tsaweb.org

The TSA is a national organization of students engaged in science, technology, engineering, and mathematics. Its website includes information about competitive events, conferences and programs, learning and career resources, and more.

Source Notes

An Innovative Career

1. Quoted in Christopher McFadden, "15 Engineers Building the Tech of the Future," Interesting Engineering, August 12, 2018. https://interestingengineering.com.
2. Quoted in McFadden, "15 Engineers Building the Tech of the Future."
3. Quoted in Meghan Brown, "STEM Salaries for New Grads Steal the Show," Engineering.com, July 11, 2017. www.engineering.com.
4. Quoted in Engineers Journal, "Why I Became an Engineer: Three People Tell Their Story," March 6, 2018. www.engineersjournal.ie.

Software Development Engineer

5. Christine Movius, "Day in the Life of Software Engineer Christine Movius," Society of Women Engineers, January 17, 2018. https://alltogether.swe.org.
6. Quoted in Prajakta Hebbar, "A Day in the Life of a Machine Learning Engineer Who Wants to Keep the Planet Clean & Green," Analytics India Magazine, June 9, 2019. https://analyticsindiamag.com.
7. Movius, "Day in the Life of Software Engineer Christine Movius."

Nanotechnology Engineer

8. Quoted in Pirelli, "Small Is Beautiful: The Power of Nanotechnology Told by Those Who Work with It Every Day," 2016. www.pirelli.com.
9. Quoted in Nano Magazine, "New Process Could Make Glass Less Likely to Break," September 24, 2019. https://nano-magazine.com.
10. Quoted in Pirelli, "Small Is Beautiful."

Biomedical Engineer

11. Quoted in Georgina Torbet, "Biotech Company 3D-Prints a Miniature Human Heart from Stem Cells," Digital Trends, September 9, 2019. www.digitaltrends.com.
12. Amanda Edwards, "A Day in the Life of Biomedical Engineer Amanda Edwards," Society of Women Engineers," November 9, 2017. https://alltogether.swe.org.
13. Edwards, "A Day in the Life of Biomedical Engineer Amanda Edwards."
14. Edwards, "A Day in the Life of Biomedical Engineer Amanda Edwards."

Robotics Engineer

15. Quoted in Amazon, "The Story Behind Amazon's Next Generation Robot," March 11, 2019. https://blog.aboutamazon.com.
16. Mikell Taylor, "So You Want to Be a Robotics Engineer," Veo Robotics, April 30, 2019. www.veobot.com.
17. Taylor, "So You Want to Be a Robotics Engineer."
18. Taylor, "So You Want to Be a Robotics Engineer."
19. Quoted in Jenny Darmody, "So You Want to Be a Robotics Engineer?," Silicon Republic, March 20, 2017. www.siliconrepublic.com.

Industrial Engineer

20. Becca Lasky, "A Day in the Life of Industrial Engineer Becca Lasky," Society of Women Engineers, November 12, 2018. https://alltogether.swe.org.
21. Quoted in Array Advisors, "Here's What Goes on Inside the Mind of a Healthcare Systems Engineer," September 10, 2018. https://array-architects.com.
22. Lasky, "A Day in the Life of Industrial Engineer Becca Lasky."

Environmental Engineering Technician

23. Quoted in *U.S. News & World Report*, "Environmental Engineering Technician," 2019. https://money.usnews.com.
24. Quoted in Robin L. Flanigan, "Hot Job: Environmental Engineering Technician Fights Contamination in Rochester,"

Rochester (NY) Democrat and Chronicle, September 26, 2017. www.democratandchronicle.com.
25. Quoted in Flanigan, "Hot Job."
26. Quoted in Careers in Oil and Gas, "Day in the Life: Arianne—Environmental Technician," April 20, 2018. https://careersinoilandgas.com.
27. Quoted in Careers in Oil and Gas, "Day in the Life."

Renewable Energy Engineer

28. Quoted in Jennifer Jackson, "A Day in the Life of a Renewable Energy Engineer," ENDS Report, May 16, 2017. www.endsreport.com.
29. Quoted in Jackson, "A Day in the Life of a Renewable Energy Engineer."
30. Quoted in Essentra Components, "A Day in the Life of: A Renewable Energy Engineer," 2019. www.essentracomponents.com.
31. Quoted in Essentra Components, "A Day in the Life of: A Renewable Energy Engineer."
32. Quoted in Jan Burns, "There Is a Bright Future for Alternative-Energy Engineers," *Houston Chronicle*, May 2, 2014. www.chron.com.
33. Quoted in Jackson, "A Day in the Life of a Renewable Energy Engineer."

Modern Structural Engineer

34. Quoted in Bright Knowledge, "My Job Explained: Structural Engineer." www.brightknowledge.org.
35. Josh Raney, "So, What Does an Internship in the Field of Structural Engineering Really Look Like?," Watry Design, 2019. https://watrydesign.com.
36. Quoted in Not Just a Princess, "Being a Female Structural Engineer," 2018. www.notjustaprincess.co.uk.

Interview with a Robotics Engineer

Jeff Hyams is a robotics engineer in Pittsburgh, Pennsylvania. He works for Neya Systems, a robotics company that develops advanced unmanned systems and off-road, self-driving vehicle technologies. He has worked as a robotics engineer for more than fourteen years. He answered questions about his career by email.

Q: Why did you become a robotics engineer?
A: I have always been interested in computers and did my undergraduate degree in computer science. I went to graduate school at the time when robotics and artificial intelligence were taking off and I was lucky enough that my school was starting up a new robotics lab as part of the computer science department. I realized that robotics requires more than just a specialized knowledge of computers, but also mechanical and electrical engineering and how it all applies to the real world and joined the lab for my graduate degree.

Q: Can you describe your typical workday?
A: In my current position, I run several different projects and manage a team of engineers, so my normal workday varies greatly. Some days I will run technical meetings with my team for individual projects, and some I will spend performing technical work myself. Luckily, in my field, I still get to work on technical problems, even while spending time making sure the larger program is running smoothly and the overall system is being completed on time.

Q: What do you like most about your job?
A: Over the years, I have worked on many different projects spanning from automated driving to machine learning to autonomous

manipulators (arms). In robotics, there is never a lack of interesting and difficult problems to solve. It is incredibly satisfying to take an idea from proposal to implementation and then on to demonstration and execution in the real world. Robotics lets you get out and away from the desk and test your work on systems in the field.

Q: What do you like least about your job?
A: This is a hard question to answer because I am lucky to have a job that I enjoy that challenges me. But with any job, there are always some tedious tasks that must get done. I may have to step away from the interesting technical work to write a report, or make sure that everything is running smoothly, or just keep track of day-to-day paperwork.

Q: What personal qualities do you find most valuable for this type of work?
A: With any technically challenging field, you need a desire to work hard to solve problems. There are not always easy, obvious solutions, and as an engineer, you very often must take someone else's ideas and figure out how to make them work on a real system. Inquisitiveness combined with hard work will take you far in any engineering or scientific field.

Q: What advice do you have for students who might be interested in this career?
A: Work hard, learn everything you can, and keep asking questions. Never assume someone solved a problem already or that there isn't a better way, and never stop learning. Even after almost two decades in robotics, I am constantly reading papers and looking for new ideas.

Q: How can high school students begin preparing now for a career as a robotics engineer?

A: Robotics is a very multi-disciplinary area of work. It requires mechanical engineering, electrical/computer engineering, and software engineering. My advice would be to look for opportunities to take classes in all of the engineering disciplines and try to see which one you enjoy the most. Take programming classes (this will help you in any field in the future), join the robotics club (or start one if your school doesn't have one), and take every STEM class you can.

Other Jobs in Engineering

Aerospace engineer
Agricultural engineer
Application engineer
Architect
Automotive engineer
Cartographer
Chemical engineer
Civil engineer
Civil engineering technician
Computer hardware engineer
Drafter
Electrical engineer
Electromechanical technician
Electronics engineer
Environmental engineer
Geological engineer
Health and safety engineer
Industrial engineering technician
Landscape architect
Marine engineer
Materials engineer
Mechanical engineer
Mechanical engineering technician
Mining engineer
Naval architect
Network engineer
Nuclear engineer
Petroleum engineer
Photogrammetrist
Plastics engineer
Quality assurance engineer
Reliability engineer
Research and development engineer
Sales engineer
Security engineer
Surveying and mapping technician
Surveyor

Editor's note: The online *Occupational Outlook Handbook* of the US Department of Labor's Bureau of Labor Statistics is an excellent source of information on jobs in hundreds of career fields, including many of those listed here. The *Occupational Outlook Handbook* may be accessed online at www.bls.gov/ooh.

Index

Note: Boldface page numbers indicate illustrations.

Accreditation Board for Engineering and Technology (ABET)
  biomedical engineering programs accreditation, 26
  engineering programs accreditation, 34, 58, 66
  engineering technology programs accreditation, 51
  industrial engineering programs accreditation, 42–43
Amazon fulfillment centers, 33–34
American Academy of Environmental Engineers & Scientists, 54, 62
American Institute for Medical and Biological Engineering, 29–30
American Society for Engineering Education, 54, 70
American Society of Civil Engineers (ASCE), 70
American Society of Mechanical Engineers (ASME), 22, 62
application software developers, 14
Arc Publishing software, 9
Association for Computer Machinery, 15
AutoCAD, 32

Biolife4D, 24
Biomedical Engineering Society (BMES), 30
biomedical engineers
  accredited programs, 26
  basic facts about, 23
  earnings, 29
  educational requirements, 26–27
  employers, 29
  information sources, 29–30
  job description, 23–26, **27,** 28–29
  job outlook, 29
  personal qualities and skills, 27–28
Blender, 32
Bureau of Labor Statistics (BLS)
  earnings
    application software developers, 14
    biomedical engineers, 29
    environmental engineering technicians, 53
    industrial engineers, 45
    software developers, 13–14
    structural engineers, 69
    systems software developers, 14
  employers of environmental engineering technicians, 53
  job outlooks
    for all occupations, 22, 61
    biomedical engineers, 29
    environmental engineering technicians, 53
    industrial engineers, 45–46
    mechanical engineers, 61
    software development engineers, 14
    structural engineers, 69
  Occupational Outlook Handbook, 53, 77

Campion, Louise, 7
CareerExplorer website, 21, 37
Carnegie Mellon University, 4
Cheremeteff, Ari, 49
Como, Vita, 58
computer-aided drafting and design, 32
Computer Science Online, 15
Computer Science.org, 15

data science, described, 9
Dhariyal, Bhaskar, 11

earnings
  application software developers, 14
  biomedical engineers, 23, 29
  environmental engineering technicians, 47, 53
  industrial engineers, 39, 45
  nanotechnology engineers, 16, 21
  renewable energy engineers, 55, 61
  robotics engineers, 31, 37
  software development engineers, 8, 13–14
  STEM careers, 5
  structural engineers, 63, 69
  systems software developers, 14
Eastwell, Hannah, 59
educational requirements
  biomedical engineers, 23, 26–27
  environmental engineering technicians, 47, 50–51
  industrial engineers, 39, 42–43

nanotechnology engineers, 16, 19
renewable energy engineers, 55, 58–59
robotics engineers, 31, 34–35, 75–76
software development engineers, 8, 11–12
structural engineers, 63, 65–66
Edwards, Amanda, 25–26
employers
    biomedical engineers, 29
    environmental engineering technicians, 53
    industrial engineers, 45
    nanotechnology engineers, 21
    personal qualities and skills desirable to, **6**
    robotics engineers, 33–34, 37
    software development engineers, 13
    structural engineers, 69
engineering careers
    accredited programs, 34, 58, 66
    demand for, 5
    types of, 77
engineering technology, accredited programs, 51
environmental engineering technicians
    basic facts about, 47
    earnings, 53
    educational requirements, 50–51
    employers, 53
    information sources, 54
    job description, 47–50, 52–53
    job outlook, 53
    personal qualities and skills, 52
environmental engineers, 47–48, 49

Farrell, Ronan, 37
Fealy, Larry, 48
Ferrara, Katherine Whittaker, 24
fossil fuels, 55

green energy, 55–56

Hay Group (Korn Ferry), 5
Heatwole, Mike, 44
Hercules (H Drive), 33–34
Hyams, Jeff, 74–76

IEEE Computer Society, 15
IEEE Engineering in Medicine and Biology Society (EMBS), 30
IEEE Nanotechnology Council, 22
IEEE Robotics & Automation Society, 37–38
industrial engineers
    accredited programs, 42–43
    basic facts about, 39
    earnings, 45
    educational requirements, 42–43
    employers, 45
    information sources, 46
    job description, 39–42, **42**, 44–45

job outlook, 45–46
    personal qualities and skills, 44
Institute of Biological Engineering (IBE), 30
Institute of Industrial & Systems Engineers (IISE), 46
International Federation of Robotics, 38
invisibility cloaks for vehicles, 4
ION energy, 11

job descriptions
    biomedical engineers, 23–26, **27,** 28–29
    environmental engineering technicians, 47–50, 52–53
    industrial engineers, 39–42, **42,** 44–45
    nanotechnology engineers, 16–18, 21
    renewable energy engineers, 55, 56–58, **60,** 60–61
    robotics engineers, 31–34, 35, 36, 74–75
    software development engineers, 8–11, **10,** 12–13, 14
    structural engineers, 63–65, **67,** 68
job outlooks
    application software developers, 14
    biomedical engineers, 23, 29
    environmental engineering technicians, 47, 53
    industrial engineers, 39, 45–46
    nanotechnology engineers, 16, 21–22
    renewable energy engineers, 55, 61
    robotics engineers, 31, 37
    software development engineers, 8, 14–15
    structural engineers, 63, 69
"Job Outlook 2019" (NACE), **6**

Kiplinger's Personal Finance, 5
Korn Ferry Hay Group, 5
Kukharenko, Sofiya, 41

Lasky, Becca, 40–41
Leger, Zachary, 51
Leshuk, Tim, 20
Lewis, Conor, 56
licenses, professional engineering (PE), 43, 66

machine learning, 9, 11
machine learning engineers, 9
machines, self-repairing, 4
Majidi, Carmel, 4
manufacturing engineers, 39
materials engineers. *See* industrial engineers
Mattholie, Robert, 56–57, 59
mechanical engineers, 61
Morris, Steven, 24
Moussavi, Zahra, 28
Movius, Christine, 9–10, 13

Nano: The Magazine for Small Science, 22

nanoscale, 16, 19
nanotechnology engineers
    basic facts about, 16
    earnings, 21
    educational requirements, 19
    employers, 21
    information sources, 22
    job description, 16–18, 21
    job outlook, 21–22
    personal qualities and skills, 19–20
National Association of Colleges and Employers (NACE), **6**
National Nanotechnology Initiative (NNI), 22
National Robotics Education Foundation, 38
National Society of Professional Engineers, 46, 54, 62, 70

Occupational Outlook Handbook (BLS), 53, 77
O'Keefe, Laura, 57–58

Pajevic, Dragan, 33, 34
PayScale, 61
Peng, Bebe, 14
personal qualities and skills
    biomedical engineers, 23, 27–28
    desirable to employers, **6**
    environmental engineering technicians, 47, 52
    industrial engineers, 39, 44
    nanotechnology engineers, 16, 19–20
    renewable energy engineers, 55, 59–60
    robotics engineers, 31, 35–36, 75
    software development engineers, 8, 12
    structural engineers, 63, 66–68
Pirelli, 18
process engineers. *See* industrial engineers
professional engineering (PE) licenses, 43, 66
Proulx, Joe, 48

Raney, Josh, 64, 66
renewable energy, 55–56
renewable energy engineers
    basic facts about, 55
    earnings, 61
    educational requirements, 58–59
    information sources, 62
    job description, 55, 56–58, **60,** 60–61
    job outlook, 61
    personal qualities and skills, 59–60
Rensselaer Polytechnic Institute, 18
Robotic Industries Association, 38
robotics engineers
    basic facts about, 31
    earnings, 37
    educational requirements, 34–35, 75–76
    employers, 33–34, 37
    information sources, 37–38
    job description, 31–34, 35, 36, 74–75
    job outlook, 37
    personal qualities and skills, 35–36, 75

self-healing materials, 4
Sharma, Ranjana, 13
Shi, Yunfeng, 18
Silvoy, Laura, 41–42
skills. *See* personal qualities and skills
Society of Manufacturing Engineers (SME), 46
soft electronics, 4
software code, 8–9
software development engineers
    basic facts about, 8
    earnings, 13–14
    educational requirements, 11–12
    employers, 13
    information sources, 15
    job description, 8–11, **10,** 12–13, 14
    job outlook, 14–15
    personal qualities and skills, 12
STEM careers, 5
structural engineering (SE) exams, 66
structural engineers
    basic facts about, 63
    earnings, 69
    educational requirements, 65–66
    employers, 69
    information sources, 70
    job description, 63–65, **67,** 68
    job outlook, 69
    personal qualities and skills, 66–68
Susanna, Antonio
    on nanomaterials as more effective use of materials, 18, 19
    on working conditions of nanotechnology engineers, 21
Symons, Katie, 65
systems software developers, 14

Taylor, Areeya, 32
Taylor, Mikell, 34, 36
Technology Student Association (TSA), 46, 54, 62, 70
3-D printers, 23–24, 32
Toyota, invisibility cloaks for vehicles, 4
training. *See* educational requirements

vehicles, invisibility cloaks for, 4

Washington Post (newspaper), 9–10
Wesselkamper, Bob, 5

Zhang, Minjuan, 4

CENTRAL ISLIP PUBLIC LIBRARY

3 1800 00357 4619

NOV 0 5 2020 3574619

Central Islip Public Library
33 Hawthorne Avenue
Central Islip, NY 11722